Quantum Coalescent Cosmos

Fractal Cosmos IV

Prime Mover Roles
An Eternal Cosmos
The Parent Universe

Stephen Blaha Ph. D.
Blaha Research

MMXXV

Copyright © 2025 by Stephen Blaha. All Rights Reserved.

This document is protected under copyright laws and international copyright conventions. No part of this book may be reproduced, stored in a retrieval system, or transmitted by any means in any form, electronic, mechanical, photocopying, recording, or as a rewritten passage(s), or otherwise, without the express prior written permission of Blaha Research.

.

ISBN: 979-8-9986071-7-2

This document is provided "as is" without a warranty of any kind, either implied or expressed, including, but not limited to, implied warranties of fitness for a particular purpose, merchantability, or non-infringement. This document may contain typographic errors or technical inaccuracies. This book is printed on acid free paper.

Rev. 00/00/01 November 16, 2025

To Margaret

Some Other Books by Stephen Blaha

SuperCivilizations: Civilizations as Superorganisms (McMann-Fisher Publishing, Auburn, NH, 2010)

Unification of God Theory and Unified SuperStandard Model THIRD EDITION (Pingree Hill Publishing, Auburn, NH, 2018).

The Exact QED Calculation of the Fine Structure Constant Implies ALL 4D Universes have the Same Physics/Life Prospects (Pingree Hill Publishing, Auburn, NH, 2019).

Passing Through Nature to Eternity ProtoCosmos, HyperCosmos, Unified SuperStandard Theory (Pingree Hill Publishing, Auburn, NH, 2022).

HyperCosmos Fractionation and Fundamental Reference Frame Based Unification: Particle Inner Space Basis of Parton and Dual Resonance Models (Pingree Hill Publishing, Auburn, NH, 2022).

The Cosmic Panorama: ProtoCosmos, HyperCosmos, Unified SuperStandard Theory (UST) Derivation (Pingree Hill Publishing, Auburn, NH, 2022).

God and and Cosmos Theory (Pingree Hill Publishing, Auburn, NH, 2023).

Newton's Apple is Now The Fermion (Pingree Hill Publishing, Auburn, NH, 2023).

Cosmos Theory: The Sub-Particle Gambol Model (Pingree Hill Publishing, Auburn, NH, 2023).

Cosmos-Universe-Particle-Gambol Theory (Pingree Hill Publishing, Auburn, NH, 2024).

Structure and Dynamics of Cosmos Theory and the Unified SuperStandard Theory (Pingree Hill Publishing, Auburn, NH, 2024).

Black Holes, White Holes, and Superluminal Starship (Pingree Hill Publishing, Auburn, NH, 2025).

The Sequence Theory Foundation of Cosmos Theory and Gambol Theory (Pingree Hill Publishing, Auburn, NH, 2025).

BEYOND TIME AND SPACE Fermion Strata & Consistency Conditions for Sequences, Cosmos Spaces and Gambols (Pingree Hill Publishing, Auburn, NH, 2025).

The C o s m o s as a Physics Organism (Blaha Research, Auburn, NH, 2025)

The Root of Reality (Blaha Research, Auburn, NH, 2025)

Fractal Cosmos (Blaha Research, Auburn, NH, 2025)

Fractal Cosmos II (Blaha Research, Auburn, NH, 2025)

Fractal Cosmos III (Blaha Research, Auburn, NH, 2025)

Available on bn.com, Amazon.com, Amazon.co.uk and other international web sites as well as at better bookstores.

CONTENTS

INTRODUCTION 1

1. THE STRUCTURE OF THE COSMOS 3

1.1 COSMOS STRUCTURE 4
1.2 COMPONENTS OF THE COSMOS 4

2. FEATURES OF THE COSMOS COMPONENTS 7

2.1 THE TWO DIMENSION FRACTAL COSMOS CURVE 7
2.2 MULTIDIMENSIONAL COSMOS FRACTAL CURVE AND HYPERSPHERES 8
2.2.1 The Hyperspheres of the Cosmos Spaces 10
2.3 COSMOS SPACE AND HYPERCOSMOS SPACES 10
2.4 THE MASTER EQUATION 11
2.4.1 Wave Function Solutions 12
2.4.2 Eigenvalue Spectrum 12
2.5 THE SIMULTANEITY OF THE FOUR COSMOS COMPONENTS 13

3. COMPONENT EQUATIONS OF THE COSMOS 17

3.1 QUANTUM COSMOS THEORY FIELD EQUATIONS 18
3.1.1 Master Equation Part 19
3.1.2 Cosmos Fractal Curve Step Lengths Part 22
3.1.3 Cosmos Spaces Part 24

4. COMPONENTS OF THE COMPONENTS 27

4.1 UNIFICATION OF THE DYNAMIC EQUATIONS' COMPONENTS OF THE COSMOS 27
4.2 RECONCILIATION OF THE COMPONENTS' DYNAMIC EQUATIONS 27

5. COALESCENCE OF THE COSMOS QUANTUM FIELD REPRESENTATIONS 29

5.1 TOTAL COALESCED FIELD Ψ 29
5.2 TOTAL COALESCED FIELD STATES 31
5.3 TIMELESS FORM OF THE TOTALLY COALESCED FIELD AND STATES 33
5.3.1 Why the Existence of Time Necessitates a Universe 33
5.4 PARTLY COALESCED FIELD 34
5.5 PARTLY COALESCED FIELD STATES 35

6. EXTENSION TO INCLUDE THE PARENT UNIVERSE 39

6.1 COSMOS SPACE – PARENT UNIVERSE QUANTUM FIELD IN R DIMENSIONS 39
6.2 PARENT UNIVERSE 40
6.2.1 Mixed Parent Universe Field 40
6.3 QUANTUM CONSISTENCY CONDITION FOR THE NUMBER OF HYPERCOSMOS SPACES AND FOR A PARENT UNIVERSE 41
6.4 QUANTUM CONSISTENCY CONDITION 41
6.5 BALANCE BETWEEN EXPANSION AND CONTRACTION 42
6.6 NUMBER OF FUNDAMENTAL PARTICLE SPECIES 43
6.7 CONDITIONS AT THE CONSISTENCY POINT 44
6.8 THERMODYNAMIC OUTWARD PRESSURE 44
6.9 CASIMIR EFFECT CONFINING FORCE 44
6.10 H BALANCE 45

6.11 N_{Cosmos} AND H' AS FUNCTIONS OF DIMENSION R AND Θ 46
6.12 WHY A PARENT UNIVERSE? 47
6.12.1 Cosmos Structure Coordinates *47*
6.12.2. Cosmos Structure Featiures *48*
6.12.3. Parent Universe Quantum Field *50*

7. HYBRID SUBUNIVERSES 51

7.1 INTERPRETATION OF THE HYBRID UNIVERSE QUANTUM STATE 51
7.2 FORM OF A HYBRID DIMENSION ARRAY 51

8. THE ETERNAL COSMOS 55

REFERENCES 57

INDEX 63

ABOUT THE AUTHOR 65

FIGURES and TABLES

Figure 1.1. The components of Cosmos Theory. The existence of HyperCosmos and 2nd Kind HyperCosmos spaces causes pairings. The constants s_1, s_2, and s_3 are overall factors for each sequence. The fourth component is the set of wave functions $\psi_{n\varphi n}(u)$.. 6
Figure 2.1. The HyperCosmos space spectrum. 14
Figure 2.2. The Cosmos Theory Limos spaces spectrum for Gambol Theory. See Blaha (2023e). 14
Figure 2.3. Part of the HyperCosmos of the Second Kind space spectrum. 15
Figure 2.4. The Cosmos Theory 2nd Kind Limos spaces spectrum. The negative r values reflect coordinates that are over constrained. The negative dimension array values indicate the array is over constraind and cannot have fermion and symmetry group assignments. 15
Figure 2.5. A section of the Cosmos Curve showing a flattened double helix form (with the helix mapped to two dimensions.) The two threads of Cosmos spaces and space-times with the "helix" connection between strands are due to the variation in θ. The connecting curved lines form a helix when the 0 spatial dimension of the cutve is considered – an analogue of a double helix similar to DNA. The variation from θ = 0 continuously to θ = π/(2n) takes $D(2^n, \theta)$ from $D(2^{r+4})$ to $D(2^{r+3})$. This change corresponds to a change from n to n + 1 as shown above with solid curving "helix" lines (the "front" view of the helix). The variation from θ = π/(2n) continuously to θ = 0 takes $D(2^n, \theta)$ from $D(2^{r+3})$ to $D(2^{r+4})$. This change corresponds to a change from n to n + 1 as shown above with broken curving "helix" lines (the "rear" view of the helix). The curved lines are not drawn to scale. 16
Figure 3.1. Extended Cosmos Theory Sequences with sequence F determined by setting λ = 0.303, which is the value of the QED couping constant e which determines the Fine Structure Constant α. The first row is for the case of a multiplier √2. Its n_φ value is an extension of the series of n_φ values 2, 4, 8 of rows A1 through B. The first and second rows are for the cases $n_\varphi = 6$ and $n_\varphi = 8$ which also may be of interest. The second row may correspond to 1.58-bit Large Language Models. Note: $32 \cdot 32 = 32^2$ and $32 \cdot 32 \cdot 32 = 32^3$ above. Note the "Calculated $\lambda_{n\varphi}$ " values are determined from the $g_{n\varphi}$ values by $\lambda_{n\varphi} = \pi^2/(\ln g_{n\varphi})^2 + ¼$. 22
Figure 6.1. Schematic diagram of virtual universe's "vacuum" bubble showing the confining Casimir force of the exterior vacuum. This force is countered by internal fermion gas pressure symbolized by a jagged interior star. 43
Figure 7.1. The type 1 and the expanded type 2 dimension arrays for <u>our</u> 4 dimension universe. The 256 dimension array has Normal and Dark sectors of 128 dimensions. We choose the right array to have a 128 dimension Normal sector and an expansion part of 128 dimensions filled with zeros. 52

Introduction

This book aims to place the features of Cosmos Theory in a tight coalescent framework where the parts follow directly from the Cosmos Fractal Curve without external intervention. This description, being more or less complete, could be described as a simulation[1] to the extent that it mirrors Reality.

The task of this book is to unite the four components of Cosmos Theory using a Quantum unification approach. It has become clear that these components reflect the same Reality when we consider the dynamics of the components in the self-generated two plus one dimension Cosmos space. We have found the unified components are independent of the time in any universe. Timeless! They furnish the basis for the determination of a Parent universe(s) that contains the subuniverses of the Cosmos.

The Parent universe is itself timeless too. A Consistency Condition exists that fixes the dimension of the largest Physical Cosmos space, the Parent universe space (a HyperCosmos space) to be 18. The spectrum of Cosmos Theory spaces then has ten HyperCosmos spaces and ten 2nd Kind HyperCosmos spaces.

The role of the Consistency Condition, which determines the number of Physical spaces, is based on the existence of a timeless Parent universe. The number of *timeless* components of Cosmos Theory is five. They are eternal. Although we describe Cosmos Theory in a "time ordered" sequence, the components, and their interrelations, are instantaneously "implemented." When does this happen? In the absence of a time it happens always – eternal instantiation.

The structure of the five component Cosmos Theory is not subject to interactions of any kind. They are dynamically determined by the Lagrangian and extended Master Equation. Thus they are stable and unchanging.

We separated the structure of Cosmos Theory from mass-energy entities. This separation enabled us to specify a stable, unchanging structure while allowing the dynamic evolution of mass-energy entities – subuniverses. We quantized subuniverses. The possibility of many subuniverses is analogous to Bruno's 16th Century Many Worlds Theory. This book extends the Cosmos spaces to have a hypersphere for each Cosmos space – "the Music of the Spaces" – Compare this to Kepler's Music of the Spheres.

Within subuniverses, particles and symmetries exist that determine their evolution. In the case of our universe we developed a Unified SuperStandard Theory over a period of over ten years that describes Elementary Particles. It is a major outgrowth of The Standard Model.

We now have a Quantum Unified Cosmos Theory. The theory emerges from a mathematical point – evolves to a one dimension line – that is used to construct a space-

[1] We say simulation, not Reality, because a modification of Cosmos Theory does not change Reality. Reality is independent of Cosmos Theory.

filled two dimension square – that becomes Cosmos space when we adjoin a “time” dimension that is simply the running length of the Cosmos Fractal Curve. In this Cosmos space we determine the five eternal components. Cosmos Theory begins with the Cosmos Fractal Curve!

Based on this development we have dimension arrays specifying the fundamental fermion spectrums, the symmetries, and the consequent interactions of particles and energies. The interactions, fermion masses and coupling constants follow from sequences specified by the Master Equation.

The thrust of this book is a Quantum weld of the components, and the constructs they generate (subuniverses and their contents); into a stream that does not necessarily require external intervention through an agency such as a Prime Mover.

The question of “Why” is partially answered. Some remaining Prime Mover issues are: the origin (the “spark”) of the creation of the Cosmos Fractal Curve – the source of all – and the origin of the hierarchy of subuniverses of the Parent universe.

1. The Structure of the Cosmos

There is a unity in the Cosmos that we are trying to elucidate. We see a structure that exists eternally since there is no time outside of universes. We see a parent universe that creates subuniverses that contain mass and energy. We have developed a description of the Cosmos of four tightly knit components that are fully unified.

In defining Cosmos Theory we have necessarily used a sequential "time-like" ordering of description. The reality is quite different. All components of Cosmos Theory structure are simultaneously present, and completed, always. No beginning and no end. No progression of developing structural features. In this respect there is no evolution of structure features except for an "initial" jump although there is no beginning. The jump is instantaneous and always – ever present.

In the description of universes we began with a Parent universe(s) that evolves dynamically creating subuniverses in its evolution. Parent universe time exists. Subuniverses have their own time. Subuniverses can evolve in their own time.

Perhaps the great issue, to which this work leads, is the source of the four components and the evolution of universes. It would be easy to say they result from the actions of a Prime Mover. However that would take the origin of the theory out of Physics. So we have sought a unifying coalescence of Cosmos Theory features that reduces (or eliminates) the need of a Prime Mover.

We suggest Quantum Theory applied to Cosmos Theory removes or largely eliminates a need for a Prime Mover. To that end we consider Quantum Cosmos Theory in detail – extending our previous books on Cosmos quantization. We believe a cascade of quantum transitions happening[2] "instantaneously," accounts for the four Cosmos components and for the creation of the Parent universe. The evolution of the Parent universes, and subuniverses, then happens dynamically according to the laws of dynamics.

The great remaining question is the reason/purpose of the Quantum Cosmos:

wHY?

to which we have no present Physical answer.

The Quantum Cosmos that we envision is the counterpart of the material universe within which we live. The universe is hierarchical with a gradation of layers of matter and interactions forming a type of sequence in terms of size and material content. The Cosmos is also hierarchical with sequences of space structures ranging from small to large dimension arrays. There is a similarity in Quantum characterization with the

[22] A situation comparable to explosive jumps in earthly evolution of species.

material, and the Cosmic, both having quantum features (seen in our previous books and later in this book.) Our universe may be said to mirror the Cosmos – both worlds reflecting a common overall structuring.

1.1 Cosmos Structure

The structure of the Cosmos does not change. It endlessly persists since it enjoys no time. The subuniverses of the Cosmos, excepting the Parent universe(s), exist within the time of the Parent universe, and within their own time as well.

The structure of the Cosmos can be viewed as the result of four component parts that have similar behavior but differ in their roles. Fig. 1.1 shows the four parts. Each component consists of two strands due to the independent existence of a possible HyperCosmos space and a Second Kind HyperCosmos space of the same space-time dimension. These spaces differ when used to create a universe. In the first case the universe has both a Normal and Dark sector. In the second case the universe only has a Normal sector similar to the Normal sector of the first case. (Normal means contents that are currently observable – at least in principle.)

1.2 Components of the Cosmos

The four components of the Cosmos which are listed in Fig. 1.1 have similar features. They all are sequences that are numbered from $n = -\infty$ to $n = +\infty$. The elements in the sequences are powers of 2: 2^n times an overall constant s_i for $i = 1, 2, 3, 4$.

The components are:

1. The steps in the construction of the Fractal Cosmos Curve using piecewise linear line segments of length 2^n.

2. The spaces in the Cosmos spectrum of spaces: One strand is strung with the HyperCosmos spaces and the other strand is strung with the Second kind HyperCosmos spaces. In the case of even n we set $n = r + 4$, resulting in HyperCosmos spaces strand, where r is the dimension of the space. In the case of odd n we set $n = r + 3$, resulting in the 2nd Kind HyperCosmos spaces strand. Note the dimension r of all Cosmos spaces is even.

3. The Master Equation energy spectrum for both HyperCosmos and 2nd Kind HyperCosmos spaces, has the form of $s_3 2^n$.

4. Each Master Equation wave function solution for each space has a zero at $s_4 2^n$ for both HyperCosmos and 2nd Kind HyperCosmos spaces.

The close semblance of these component sequences implies that they are representations of the same Physics. As a result we find the Fractal Cosmos Curve embodies the Physics of the structure of Cosmos Theory. The introduction of a Quantum Cosmos Theory formulation further brings out the close relationship and the primary role of the Fractal Cosmos Curve.

As a result we obtain a tightly bound Quantum structure for Cosmos Theory. Note that Cosmos Theory structure has no mass-energy and therefore is solely a concept – a structure of spaces. One might compare Cosmos Theory to Plato's Theory of Ideals – a theory that posits every material thing has a corresponding abstract Ideal which is the perfection of its material counterpart.

The Cosmos is also hierarchical with sequences of space structures ranging from small to large dimension arrays. There is a similarity between the Material Universe of our experience and the Cosmos. They both are hierarchical. The Material Universe has a distinct hierarchy of forces and matter – ranging from the small to the large. The Cosmos similarly has a hierarchy of spaces. Both are quantum.

The universe may be said to echo the Cosmos – both "worlds" have a similar overall structuring.

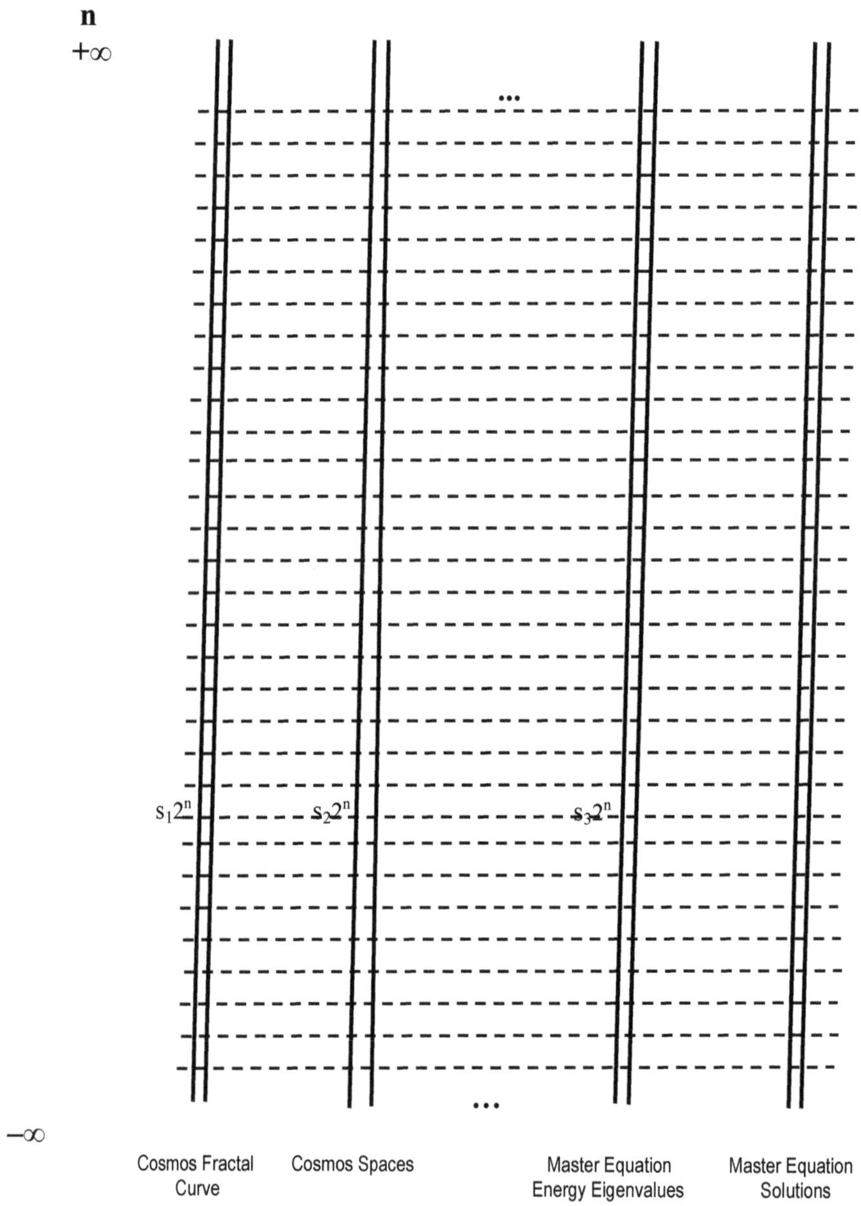

Figure 1.1. The components of Cosmos Theory. The existence of HyperCosmos and 2nd Kind HyperCosmos spaces causes pairings. The constants s_1, s_2, and s_3 are overall factors for each sequence. The fourth component is the set of wave functions $\psi_{n\varphi n}(u)$.

2. Features of the Cosmos Components

We now examine the features of Cosmos structure.

2.1 The Two Dimension Fractal Cosmos Curve

In previous work we distinguished between the growth of the Cosmos from a point to a piecewise linear unit line segment, and the construction of the fractal curve from piecewise linear line segments to fill a two dimension unit square (followed by an extension to a continuous two dimension space.)

We now join these two constructions to form a fractal curve generated from a zero dimension point to a two dimension unit square.

The construction of the unit line segment follows by a series of steps from infinitesimal line segments up to a segment of length ½ based on the identity:

$$\sum_{k=\infty}^{1} 2^{-k} = 1 \qquad (2.1)$$

The generated unit line, which is fractal, has the Hausdorff fractal dimension 1 by the fractal relation:

$$D = \ln(N)/\ln(\varepsilon) = 1 \qquad (2.2)$$

with $N = 2$ = the number of copies of the shape formed from each copy of the previous iteration of the curve under construction, and $\varepsilon = 2$ = the number of copies are scaled downward from the previous step. (Each step length is twice the previous step length.)

To the unit length 1 line we adjoin steps starting from a piece-wise linear line segment of length 1 created previously to create the rest of a space filling curve that fills a unit square. Then the curve is extended to be continuous and scaled upward to a two dimension space, which we call the Cosmos Space.

The Hausdorff dimension for the construction of the space filling curve from line segments of length 1 through infinity is

$$D = 2 \qquad (2.3)$$

since $N = 4$ and $\varepsilon = 2$ for this fractal curve part of the Cosmos Fractal Curve.

The difference between the N values for the unit line segment construction lies in the nature of the construction process. The $N = 2$ value is due to the construction of line segments in one linear dimension to create a unit line. At each stage of the construction the line segment doubles. The $N = 4$ value of the construction from a line segment of length 1 to a full fractal is due to the construction of line segments in *two* linear dimensions with the line segments broken to fit in a unit square. As a result the N value grows by a factor for four copies of $N = 2^2 = 4$ leading to Hausdorff dimension 2.

The Cosmos Fractal Curve has two parts: a D = 1 part and a D = 2 part that lead to a growth from a point to a two dimension unit square.

The points of the Fractal curve can be continuously connected to form a filled unit square that can then be scaled to a two dimension infinite space.

2.2 MultiDimensional Cosmos Fractal Curve and Hyperspheres

The above discussion of the two dimension Cosmos Fractal Curve can be generalized to the case of a multidimension Fractal Curve. This generalization would enable us to define a generalized Master Equation that ties its eigenvalues directly to the two dimension Cosmos Master Equation Space.

Consider a d dimension hypersphere (hypercube). We will define a construction process that takes a one dimension line segment as discussed above to a d dimension unit hypersphere (and then to a higher dimension space.). For example, we constructed the 2 dimension unit square using construction steps that had steps with piecewise linear lengths growing by factors of 2. This choice corresponded to N = 4 in eq. 2.2.

We now choose steps with piecewise linear lengths growing by factors of

$$N = 2^{d-1} \tag{2.4}$$

to construct a unit d dimension hypersphere. Then the generated hypersphere fractal dimension is

$$D = \ln(2^{d-1})/\ln(\varepsilon) = d - 1 \tag{2.5}$$

Note: N = the number of copies of the shape formed from each copy of the previous iteration of the curve under construction. In this case there are 2^{d-1} copies of the shape that is formed from each copy of the previous iteration of the curve under construction.

With the choice of $N = 2^{d-1}$ we obtain a sequence of values of piecewise linear line copies of length 2^{d-1}. For d = 2 we obtain a two dimension fractal space-filling unit square that becomes the two dimension Cosmos Master Equation Space.

For choices of d we find:

d	Unit Hypersphere Dimension
0	Zero dimension point
1	One dimension unit line
2	Two dimension unit square
3	Three dimension unit cube
…	
k	k dimension unit hypersphere

For each d value we can visualize the construction process as folding each line segment for each construction step to fit in the unit "hypersphere." For d = 0 the folding collapses the constructed curve to a point. For d = 1 the folding collapses the constructed curve to a unit line.

Previously, we specified the lengths for the two dimension unit square as 2^n where n is the number of the construction step. In the present case we have lengths of

$$(2^{d-1})^n = 2^{(d-1)n} \tag{2.6}$$

Since d is an integer we can define a sequence of powers of 2^{d-1} for each construction step n.

For the set of hyperspheres, we can visualize the hypersphere fractal curves grow step by step with unit lengths of size 2^{d-1}. These lengths are divided into d lines of length $2^{d-1}/d$ that "populate" each dimension of the space-filled unit hypersphere of dimension d.[3]

We set d = 2 to obtain the two dimension Cosmos Master Equation Space with fractal step n length of size 2^n. Then the sequence of step lengths in Fig. 2.5 matches the eigenfunction sequence of the Master Equation:

$$C_{Hn\varphi n} = c_{Hn\varphi n} 2^r \tag{2.7}$$

where r is even. When n = r + 4 with $c_{Hn\varphi n} = 2^4$ we obtain HyperCosmos spaces. When n = r + 3 with $c_{Hn\varphi n} = 2^3$ we obtain 2nd Kind HyperCosmos spaces. The choice of the value of n is made to relate fractal steps to Cosmos spaces.

We can transform the hypersphere sequence to correspond to the HyperCosmos spaces sequence by using the r+4 dimension Hypersphere fractal curve with

$$D = \ln(2^{r+4})/\ln(\varepsilon) = r + 4 \tag{2.8}$$

We can transform the Cosmos spaces sequence into a hypersphere sequence to correspond to the 2nd Kind HyperCosmos spaces sequence by using the r+3 dimension Hypersphere fractal curve with

$$D = \ln(2^{r+3})/\ln(\varepsilon) = r + 3 \tag{2.9}$$

Thus we have a relation between fractal curves, and the solutions of the Master Equation, and space-time dimensions. Cosmos spaces have corresponding fractal curves that grow to space filled hyperspheres.

---

[3] The map discussed here to a hypersphere is the same map previously described by the author as an inverse map of a one dimension fractal to a Z-order Lebesgue d dimension space filling curve (Morton space-filling curve) and thence to a d dimension universe (space).

2.2.1 The Hyperspheres of the Cosmos Spaces

The space-time dimensions of the Cosmos spaces have corresponding Hyperspheres. For a HyperCosmos space of space-time dimension r, the corresponding hypersphere has dimension r + 4. For the space of our universe, if it is a HyperCosmos space, the hypersphere has 8 dimensions.

For a 2nd Kind HyperCosmos space of space-time dimension r, the corresponding hypersphere has dimension r + 3. For the space of our universe, if it is a 2nd Kind HyperCosmos space, the hypersphere has 7 dimensions.

Historically there is a remarkable analogy between the Cosmos hyperspheres and Kepler's use of regular polyhedra to understand the periodic conjunctions of planets. For the six planets known at his time he suggested that a nested set of spherical orbs, with each containing a regular polyhedron could yield the six layers of planets. The polyhedrons were octahedron, icosahedron, dodecahedron, tetrahedron and cube. He related the size of a planet's orbit to the length of its orbital period.

A surprising prefigurement of the Cosmos hyperspheres! Now we have *The Music of the Hyperspheres*!

2.3 Cosmos Space and HyperCosmos Spaces

The Cosmos Fractal Curve is used to generate a space called a *Cosmos Space* that has two spatial dimensions and one time like dimension. We will discuss details of this space in a later chapter. The Cosmos Space, so generated, is simultaneous in construction with the other Cosmos components due to the absence of time. Thus we may use it as the space of the Master Equation and its solutions as we do.

The Fractal Cosmos Curve has several roles. Its unit square and its extension to a two dimension space becomes the Cosmos Space. The distance along the curve is used as a time coordinate in Cosmos Space, and in the Master Equation in particular. Although the Cosmos Fractal Curve has discontinuous derivatives, the distance along the curve is continuous.

We can associate the steps of the Cosmos Fractal Curve with HyperCosmos spaces. HyperCosmos spaces are defined to have a dimension array of size

$$2^{r+4} \quad \text{HyperCosmos Spaces} \qquad (2.10)$$
$$2^{r+3} \quad 2^{nd}\ \text{Kind HyperCosmos Spaces}$$

where r is the dimension of the space-time within the dimension array for each space.[4] Dimension arrays are sets of dimensions that play the role of place holders that are used to define fundamental fermion spectrums and symmetry group irreducible representations.

[4] The "3" and "4"appearing in the definition of the dimension array for each space is due to the map between dimension arrays and the creation/annihilation operators of Fourier expansions of fermion fields. This map is discussed in our previous books in some detail. We use PseudoQuantum fermion fields in those discussions.

Corresponding to the set of HyperCosmos spaces there is a set of spaces with negative dimension that we have called Limos spaces. These spaces correspond to the unit line segment values in eq. 2.1: 2^{-k}. Fig. 2.2 shows the Limos HyperCosmos spaces.

The HyperCosmos and 2nd Kind HyperCosmos spaces correspond to steps in the construction of the D = 2 part of the Cosmos Curve. See Fig. 2.5.

The HyperCosmos spaces appear in Figs. 2.1 – 2.5. The spaces are specified in Fig. 2.5 with an index n that corresponds to either HyperCosmos or 2nd Kind HyperCosmos spaces. The dimension[5] r is necessarily even to avoid double counting of dimension arrays and their spaces. For even n

$$n = r + 4 \tag{2.11}$$

for HyperCosmos and Limos spaces.

For odd n

$$n = r + 3 \tag{2.12}$$

for 2nd Kind HyperCosmos spaces and Limos spaces.

These considerations lead to the Fractal Cosmos Curve associated with Cosmos spaces shown in Fig. 2.5. Note the sets of spaces correspond to n values ranging from n = $-\infty$ to n $-$ $+\infty$.

2.4 The Master Equation

At this point we have a Fractal Cosmos Curve and a corresponding set of spaces for each step in the construction of the curve. With no time present the curve and set of spaces exists simultaneously with the Master Equation. The Master Equation is set in the infinite two dimension space generated from the Fractal Cosmos Curve.

We define a time coordinate in that space as measured by running length of the Fractal Cosmos Curve as it is being constructed. Thus there are 2+1 dimensions in the Cosmos Space that we use to define a Lagrangian that becomes the Master Equation.[6] The Master Equation generates an eigenvalue spectrum and a set of wave function solutions. Despite the introduced time coordinate the simultaneity of all Cosmo components remains maintained.

The Lagrangian density is

$$\mathcal{L} = ½(\partial^\mu \varphi(z)\partial_\mu\varphi(z)/m \;-\; V(z)\varphi(z)\varphi(z)) \tag{2.13}$$

[5] The space-time dimensions can be arbitrarily large until a stability (consistency) condition limits the number of dimensions. Dimensions greater than four are acceptable in Quantum Field Theories that implement our Two Tier Theory. For polynomial Lagrangians and arbitrary space-time dimensions all perturbation theory calculations are finite. In addition all perturbation theory calculations using Two Tier Theory give the known results for low momenta. See Blaha (2002).

[6] This procedure is discussed in detail in Blaha (2025h).

where z is the 3-vector (t, x, y), $\partial^\mu \varphi(z)$ is the covariant derivative, a (non-mass) constant m is introduced for use later, and V(z) is a scaling potential

$$V(z) = \lambda/[m^2(x^2 + y^2)] \tag{2.14}$$

with λ a constant, From it we derive the Master Equation:

$$[(\partial^2/\partial x^2 + \partial^2/\partial y^2)/m + \lambda/m^2r^2 + C)]\psi = 0 \tag{2.15}$$

where $r = (x^2 + y^2)^{1/2}$ after letting

$$\varphi = e^{iEt}\psi(x, y) \tag{2.16}$$

with $C = -E^2$.

2.4.1 Wave Function Solutions

The form of the complete wave function solution of the Master Equation is

$$\psi_{E,\,k,\,n_\varphi,\,n}(u, t) = e^{iEt}\exp^{ik\varphi}\psi_{n_\varphi n}(u) \tag{2.17}$$

The ψ(u) radial factor solution is

$$\psi_{n_\varphi n}(u) \approx B\cos((\lambda_{n_\varphi} - 1/4)^{1/2}\ln((-mC_{n_\varphi n})^{1/2}u) + b) \tag{2.18}$$

where c_{n_φ}, b, $C_{n_\varphi n}$, and B are constants. The integer constant n is the radial quantum number. The integer constant n_φ is an extension of the angular quantum number $n_{\varphi ang}$ based on a separation of variables where we set

$$\lambda = k^2 + n_{\varphi ang}{}^2$$

and then define

$$\lambda_{n_\varphi} = n_\varphi{}^2/m \tag{2.19}$$

for use in the Master Equation

$$(\partial^2/\partial u^2 + 1/u\ \partial/\partial u + \lambda_{n_\varphi}/u^2 + mC)\psi(u) = 0 \tag{2.20}$$

2.4.2 Eigenvalue Spectrum

The eigenvalues[7] of the Master Equation are

$$C_{n_\varphi n} = c_{n_\varphi}\exp(2\pi n/(\lambda_{n_\varphi} - 1/4)^{1/2}) \tag{2.21}$$

They form a sequence of powers of

$$g_{n_\varphi} = \exp(2\pi/(\lambda_{n_\varphi} - 1/4)^{1/2}) \tag{2.22}$$

---

[7] We have modified the $C_{n\varphi n}$ value by introducing a factor of 2 in the exponent since $C_{n\varphi n} \sim E^2$ in the Master Equation – E^2 is an energy squared term.

with eigenvalues

$$C_{n\varphi n} = c_{n\varphi}\, g_{n\varphi}{}^{n} \tag{2.23}$$

2.5 The Simultaneity of the Four Cosmos Components

All four components are simultaneous. They may be said to be ever present since there is no time in the Cosmos structure. Times only exist within universes.

As Fig. 1.1 indicates all four have the same values for each index value n:

$$2^{n} \tag{2.24}$$

up to an overall constant when $g_{n\varphi} = 2$.

1. Thus the energy (squared) eigenvalue is

$$C_{n\varphi n} = c_{n\varphi}\, g_{n\varphi}{}^{n} = c_{n\varphi}\, 2^{n} \tag{2.25}$$

The energy values are mapped to space-time dimensions.

2. The steps in the construction of the Fractal Cosmos Curve have piecewise linear line lengths of 2^n. The line length is mapped to the space-time dimension.

3. Correspondingly, the wave function $\psi_{n\varphi n}(u)$ has a zero at

$$\cos((\lambda_{n\varphi} - ¼)^{½}\ln((-mC_{n\varphi n})^{½}u) + b) = 0 \tag{2.26}$$

implying

$$(\lambda_{n\varphi} - ¼)^{½}\ln((-mC_{n\varphi n})^{½}u) + b = (n + ½)\pi \tag{2.27}$$

When $b = \pi/2$ we find[8]

$$u_n = -1/(m\, c_{n\varphi}{}^{½}) \tag{2.28}$$

Note there is a 1 to 1 correspondence of the four components.

---

88 In Blaha (2025h) we found, where we let $(-mC_{n\varphi n})^{½}u \rightarrow u$, the zero to be at

$$u_n = \exp(n\pi/(\lambda_{n\varphi} - ¼)^{½}) = C_{n,n\varphi} \tag{3.18}$$

where

$$C_{n\varphi n} = \exp(n\pi/(\lambda_{n\varphi} - ¼)^{½}) \tag{3.9}$$

Or

$$C_{n\varphi n} = 2^{n}$$

like the other components' sequence values.

THE HYPERCOSMOS SPACES SPECTRUM

Blaha Space Number N_s	Cayley-Dickson Number n_c	Cayley Number Cn d_c	Dimension Array column length d_{cd}	Dimension Array Size $d_{dN}=d_{dr}$	Space-time-Dimension r	CASe Group $su(2^{r/2},2^{r/2})$ CASe
0	10	1024	2048	2048^2	18	su(512,512)
1	9	512	1024	1024^2	16	su(256,256)
2	8	256	512	512^2	14	su(128,128)
3	7	128	256	256^2	12	su(64,64)
4	6	64	128	128^2	10	su(32,32)
5	5	32	64	64^2	8	su(16,16)
6	4	16	32	32^2	6	su(8,8)
7	**3**	**8**	**16**	$\mathbf{16^2}$	**4**	**su(4,4)**
8	2	4	8	8^2	2	su(2,2)
9	1	2	4	4^2	0	su(1,1)
10	0	1	2	2^2	-2	su(0,0)
11	-1	½	1	1	-4	su(-1,-1)

Figure 2.1. The HyperCosmos space spectrum.

THE Limos SECTOR SPACES SPECTRUM

Blaha Space Number $N = O_s$	Cayley-Dickson Number n	Cayley Number d_c	Row Size d_{cr}	Dimension Array Size $d_{dN}=d_{dr}$	Space-time-Dimension r	CASe Group $su(2^{r/2},2^{r/2})$ CASe
10	0	1	2	2^2	-2	U(1)
11	-1	½	1	1^2	-4	U(½)
12	-2	¼	½	$½^2$	-6	U(¼)
13	-3	1/8	¼	$¼^2$	-8	U(1/8)
14	-4	1/16	1/8	$1/8^2$	-10	U(1/16)
				•		
				•		
				•		
∞	−∞	0	0	0	−∞	U(0)

Figure 2.2. The Cosmos Theory Limos spaces spectrum for Gambol Theory. See Blaha (2023e).

HYPERCOSMOS OF THE SECOND KIND SPACES SPECTRUM

Blaha Space Number $N = O_s$	Cayley-Dickson Number n	Cayley Number d_c	Dimension Array size $d_{dN2}=d_{dr2}$	Space-time-Dimension r	CASe Group $su(2^{r/2},2^{r/2})$ CASe
0	10	1024	1024×2048	18	su(512,512)
1	9	512	512×1024	16	su(256,256)
2	8	256	256×512	14	su(128,128)
3	7	128	128×256	12	su(64,64)
4	6	64	64×128	10	su(32,32)
5	5	32	32×64	8	su(16,16)
6	4	16	16×32	6	su(8,8)
7	**3**	**8**	**8×16**	**4**	**su(4,4)**
8	**2**	4	4×8	2	su(2,2)
9	**1**	2	2×4	0	su(1,1)

Figure 2.3. Part of the HyperCosmos of the Second Kind space spectrum.

THE 2nd Kind Limos SECTOR SPACES SPECTRUM

Blaha Space Number $N = O_s$	Cayley-Dickson Number n	Cayley Number d_c	Row Size d_{cr}	Dimension Array Size $d_{dN}=d_{dr}$	Space-time-Dimension r	CASe Group $su(2^{r/2},2^{r/2})$ CASe
: 10	0	1	2	2	-2	U(1)
11	-1	½	1	0	-4	U(½)
12	-2	¼	½	$-½^2$	-6	U(¼)
13	-3	1/8	¼	-3/16	-8	U(1/8)
14	-4	1/16	1/8	-7/64	-10	U(1/16)
				•		
				•		
				•		
∞	$-\infty$	0	0	0	$-\infty$	U(0)

Figure 2.4. The Cosmos Theory 2nd Kind Limos spaces spectrum. The negative r values reflect coordinates that are over constrained. The negative dimension array values indicate the array is over constraind and cannot have fermion and symmetry group assignments.

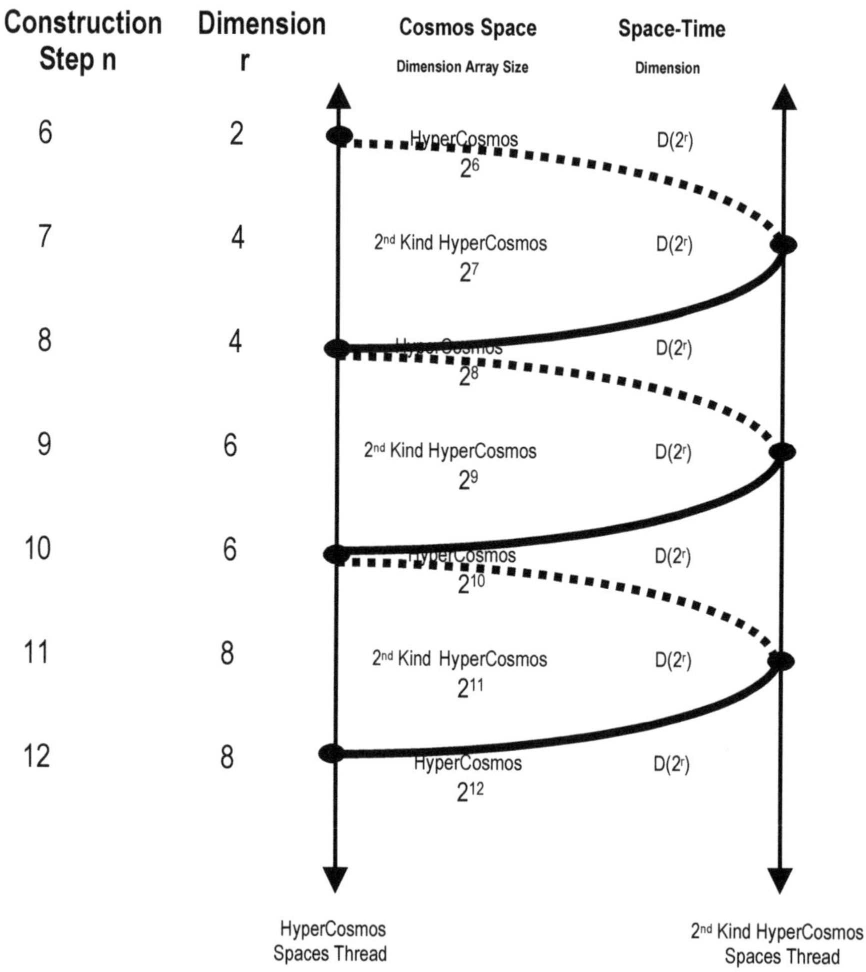

Figure 2.5. A section of the Cosmos Curve showing a flattened double helix form (with the helix mapped to two dimensions.) The two threads of Cosmos spaces and space-times with the "helix" connection between strands are due to the variation in θ. The connecting curved lines form a helix when the 0 spatial dimension of the cutve is considered – an analogue of a double helix similar to DNA. The variation from $\theta = 0$ continuously to $\theta = \pi/(2n)$ takes $D(2^n, \theta)$ from $D(2^{r+4})$ to $D(2^{r+3})$. This change corresponds to a change from n to n + 1 as shown above with solid curving "helix" lines (the "front" view of the helix). The variation from $\theta = \pi/(2n)$ continuously to $\theta = 0$ takes $D(2^n, \theta)$ from $D(2^{r+3})$ to $D(2^{r+4})$. This change corresponds to a change from n to n + 1 as shown above with broken curving "helix" lines (the "rear" view of the helix). The curved lines are not drawn to scale.

3. Component Equations of the Cosmos

The four components of the Cosmos can be separately determined. Yet we have seen they lead to similar sequences as in Fig. 1.1. We now seek to unite them in a common framework by introducing Quantum Theory. In doing this extension we realize there is no apparent Observer similar to the Observer in Quantum Mechanics. However we view the existence of an Observer as not required.[9] On this basis we proceed to Second Quantize Cosmos Theory to create a coalescent Cosmos.

The coalescence of the features of the Quantum Cosmos Theory enables the theory to "evolve" from component to component seamlessly without the introduction of a Prime Mover executing steps of the evolution.[10] When we say "evolution" we view it as instantaneous and simultaneous and independent of any time. It is perhaps analogous to jumps in evolution of species seen here on earth. We see the creation of universes as an evolutionary process.

The Cosmos will now have a structure specified by Quantum Cosmos Theory. We view all components of the structure of the theory as simultaneously existent having no time ordering. In particular, the two dimension space, within which the Master Equation is operative, has quantum components: the Fractal Cosmos Curve, the set of Cosmos spaces, and the solutions of the Master Equation. To understand these components and their features we have introduced a pseudo-time for the two dimension space and proceeded to follow a "path" of evolution for their origin.

We treat all components as coinciding for all "time" *if* there were a time. The space, which we call the Cosmos space, has two spatial coordinates and one pseudotime coordinate whose sole role is to support the Master Equation.

We define a scalar quantum field theory with the dynamics of the Master Equation. Its solutions are second quantized. We specify one Mathematical point as the

---

[9] Those who may wish to invoke a Prime Mover as the Observer may do so. However a Prime Mover that only seeks to "start" the Cosmos and does not act further in its evolution is beyond Physics and thus not a subject of discussion here. Our choice of a coalescent Quantum formulation of the Cosmos avoids the issue of specifying the linkup of the components of the Cosmos. It does not affect the question of an initial Prime Mover.

[10] The possible roles of a Prime Mover would include the initiation of an ever present Cosmos, the creation of universes, and possibly the Quantum "tweaking" of events – external and internal to universes. A Prime Mover (God) can perform miracles and change history using Quantum "tweaks" in Quantum processes. With Quantum chance in physical processes, it becomes possible for a Prime Mover (God) to "tweak" an event quantum "mechanically" by choosing a certain one of its probable outcomes. Then the tweaked process can cascade to make a certain final result happen. The tweak(s) does not affect the Physics any more than adding or removing a water drop affects an ocean.

Thus one sees the possibility of the external influence on events using Quantum uncertainty *without violations of Quantum Physical dynamic equations.* With this view one might conclude that *Quantum* evolution is required for a Prime Mover (God) to gently influence the course of events. See Blaha (2018e).

origin point of the Fractal Cosmos Curve which is also quantized step by step in its construction. This Quantum Fractal Cosmos Curve can be specified to consist of the sequence of Master Equation solutions, and to have a set of HyperCosmos Spaces that are also quantized. The Fractal Cosmos Curve and the Master Equation solutions *coexist* for all "time" meaning *always*.

Each solution in the set of second quantized Master Equation solutions has an energy. The energies fix the dimension spaces' sizes. The solutions correspond to the structure of the Quantum Fractal Cosmos Curve. Again we note that the "logical" ordering of components and their features is not the Physical Reality of this theory of coexisting components.

As a result we obtain a tightly bound Quantum structure for Cosmos Theory. Note that Cosmos Theory structure has no mass-energy and therefore is solely a concept. One might compare this aspect of the theory with Plato's Theory of Ideals – a theory that posits every material thing has a corresponding abstract Ideal which is the perfection of its material counterpart.

When Cosmos quantization is introduced the HyperCosmos and Second Kind HyperCosmos spaces wave functions (states) become parts of a composite space where the Normal sectors form a composite, and the Dark sector of the HyperCosmos space "stands alone."

Each of the spaces defines its dimension array and specifies the structure of corresponding universes.[11]

3.1 Quantum Cosmos Theory Field Equations

The goal of the theory is the quantum coalescence of all four components of Cosmos Theory. We assume there are three sets of quantum fields:

Quantum fields for the steps of the Cosmos Fractal Curve
Quantum fields for the set of Cosmos spaces
Quantum fields for the Master Equation solutions.

Thus we anticipate the overall Lagrangian form of the Quantum Cosmos Theory is:

$$\mathcal{L} = \mathcal{L}_{CC} + \mathcal{L}_{CS} + \mathcal{L}_{ME} \tag{3.1}$$

where $\mathcal{L}_{CC}$ is the Cosmos Curve Lagrangian, $\mathcal{L}_{CS}$ is the Cosmos spaces Lagrangian, and $\mathcal{L}_{ME}$ is the Master Equation Lagrangian, Eq. 3.1 leads to a separable dynamic equation of the form:

$$(\text{ Fractal line Part} + \text{ Spaces Part } + \text{ Master Equation Part})\psi + E\psi = 0 \tag{3.2}$$

[1111] The Parent universe(s) exists simultaneously with the four Cosmos Theory components of Fig. 1.1. Subuniverses are created within the Parent universe.

3.1.1 Master Equation Part

Upon separation of the terms we have the Master Equation Lagrangian (as used before):

$$\mathcal{L}_{ME} = ½(\partial^{\mu} \varphi(z)\partial_{\mu}\varphi(z)/m \; - V(z)\varphi(z)\varphi(z)) \tag{3.3}$$

where z is the 3-vector (t, x, y), $\partial^{\mu}\varphi(z)$ is the covariant derivative, a (non-mass) constant m is introduced for use later, and V(z) is a scaling potential

$$V(z) = \lambda/[m^2(x^2 + y^2)] \tag{3.4}$$

with λ a constant, From it we obtain the Master Equation:

$$[(\partial^2/\partial x^2 + \partial^2/\partial y^2)/m + \lambda/m^2r^2 + C_{ME})]\psi = 0 \tag{3.5}$$

where $r = (x^2 + y^2)^{½}$ after letting

$$\varphi = e^{iE_{ME}t}\psi(x, y) \tag{3.6}$$

with $C_{ME} = -E_{ME}^2$. If we use polar coordinates (r, θ):

$$\begin{aligned} x &= u\cos(\varphi) \\ y &= u\sin(\varphi) \end{aligned} \tag{3.7}$$

and set

$$\psi = \psi(u)\Phi(\varphi) \tag{3.8}$$

where

$$\Phi(\varphi) = \exp(ik\varphi) \tag{3.9}$$

we find

$$\begin{gathered} (\partial^2/\partial u^2 + 1/u\, \partial/\partial u + \partial^2/\partial\varphi^2 + \lambda/mu^2 + mC)\psi = 0 \\ (\partial^2/\partial\varphi^2)\psi = -k^2\psi \end{gathered} \tag{3.10}$$

The quantum number k must be a whole integer to avoid irregularities in ψ(r). We set $\lambda = k^2 + n_{\varphi}^2$. Substituting and separating factors in eq. 4.4 we find

$$\begin{gathered} (\partial^2/\partial u^2 + 1/u\, \partial/\partial u + (\lambda - k^2)/(mu^2) + mC)\psi(u) = 0 \\ (\partial^2/\partial u^2 + 1/u\, \partial/\partial u + \lambda_{n\varphi}/u^2 + mC)\psi(u) = 0 \end{gathered} \tag{3.11}$$

where[12] we assign the energy $E = (-mC_{n\varphi})^{½}$ based on the solution of Case and set

---

[12] The eigenvalue spectrum results from the imposition of orthogonality condition on the solution function near u = 0 according to von Neumann. The value of $-mC_{n\varphi n}$ can be taken to be positive since its sign can be changed by Wick rotations of the time variable: t → it.

$$\lambda_{n\varphi} = (\lambda - k^2)/m \tag{3.12}$$

We now define

$$\lambda_{n\varphi} = n_\varphi^2/m \tag{3.13}$$

based on a similarity to the value obtained in the semi classical Bohr Model[13] that we presented in an earlier book. The constant n_φ is a result of the φ eqns. 3.10 and 3.12 above.

The value of $\lambda_{n\varphi}$ can also be calculated from

$$\lambda_{n\varphi} = \pi^2/(\ln g_{n\varphi})^2 + 1/4 \tag{3.13a}$$
$$\lambda_{n\varphi}' = (\lambda_{n\varphi} - 1/4)^{1/2} \tag{3.13b}$$

from eq. 3.16 below.

The energy (based on Case's energy eigenvalue) is

$$E_{n\varphi n} = (-mC_{n\varphi n})^{1/2} = (-mc_{n\varphi})^{1/2}\exp(-n\pi/(\lambda_{n\varphi} - 1/4)^{1/2}) \tag{3.14}$$

where $c_{n\varphi}$ is an arbitrary constant. Upon letting[14] n → -n, we have

$$E_{n\varphi n} = s_3 \exp(n\pi/(\lambda_{n\varphi} - 1/4)^{1/2}) = s_3 g_{n\varphi}{}^n = s_3 2^n \tag{3.15}$$

using eqs 3.16 and 3.17 and the row A2 values of Fig. 3.1 below where s_3 is a constant[15] defined in Fig. 1.1 for the Master Equation eigenvalues (since the energy is only defined up to an overall constant due to the scaling), where n is a radial quantum number (related to r), and where n_φ is the angular momentum quantum number.

The growth factor (multiplier) for each power series sequence is

$$g_{n\varphi} = \exp(\pi/(\lambda_{n\varphi} - 1/4)^{1/2}) \tag{3.16}$$

The rationale for defining the Master Equation for a Quantum field is its use to define sequences of spaces, coupling constants, and fermion masses as we have done in earlier books. This feature depends solely on n_φ values. The other quantum fields are not used to define sequences of coupling constants and fermion masses.

Fig. 3.1 shows the resulting Sequence Table.[16] We have previously related the multipliers to fractals.

Note the A2 row for Cosmos spaces where $n_\varphi = 4$ and $g_g = g_{n\varphi=4} = 2$ corresponds to the sequence of energies $s_3 2^n$ of the Master Equation. The constant m was determined in Blaha (2025b) by requiring

---

[13] See section 2.3 of Blaha (2025e).

[14] The sign of n can be positive or negative due to the scaling property of the Master Equation.

[15] The constant s_3 can be positive or negative. Its sign can be changed by Wick rotations of the time variable: t → it.

[16] S. Blaha (2025f).

$$g_{n_\varphi=4} = 2 \tag{3.17}$$

$$m = 0.7696 \tag{3.17a}$$

The form of the complete Master Equation wave function solution (the fourth component) is

$$\psi_{E,\, k,\, n_\varphi,\, n}(u, t) = e^{iEt} \exp^{ik\varphi} \psi_{n_\varphi n}(u) \tag{3.18}$$

The ψ(u) radial factor solution is

$$\begin{aligned} \psi_{n_\varphi n}(u) &\approx B\cos((\lambda_{n_\varphi} - \tfrac{1}{4})^{1/2} \ln((-mC_{n_\varphi n})^{1/2} u) + b) \\ &\approx B\cos((\lambda_{n_\varphi} - \tfrac{1}{4})^{1/2} \ln(E_{n_\varphi n} u) + b) \\ &\approx B\cos((\lambda_{n_\varphi} - \tfrac{1}{4})^{1/2} \ln(2^n u) + b) \end{aligned} \tag{3.19}$$

where b and B are constants.

Thus the third and fourth components of Cosmos Theory are determined.

SEQUENCES TABLE

Sequence Fractal	n_φ	$g_{n\varphi}$ Fractal Dimension multiplier factor	$\lambda_{n\varphi}=n_\varphi^2/m$	$\lambda_{n\varphi} = \pi^2/(\ln g_{n\varphi})^2+¼$ Calculated
Fractal	12	$1,26 = \ln_3(4)$	187.1	185
Fractal	10	$1.317 = \ln_{2.7}(3.7) = \ln_5(8)$	129.9	130.4
Fractal	8	$1.41 = \ln_{2.36}(3.36) = \ln_4(7)$	83.16	83.9
Sierpinski Triangles Fractal; 1.58-bit Large Language Model	6	$1.585 = \ln_2(3)$	46.778	46.77
A1 Cosmos Fractal Curve line steps	4	$2 = \ln_2(4)$	20.79	20.79
A2 Cosmos spaces of dimension r dimension array column lengths $2^{r/2+2}$	4	$2 = \ln_2(4)$	20.79	20.79
A3 The sequence of Unification Spaces	4	$2 = \ln_2(4)$	20.79	20.79
B Coupling Constants α Multiplier	2	$4 = \ln_2(16)$	5.398	5.39
C Fermion Sequence 2 Masses	1	$2/3 \times 2^5 = \ln_2((32^{32})^{0.133})$	1.3	1.3
D Fermion Sequence 1 Masses	$½^{½}$	$3/2 \times \pi 2^5 = \ln_2((32^{32})^{0.3\pi})$	0.65	0.81
E n_φ Values Sequence; A Sequence of Sequences	½	$95939 = 5.96 \ln_2(32^{32\cdot 32\pi})$	0.325	0.325
F New n_φ Values Sequence based on α_{QED}	$0.48 \approx 2^{½}/3$	$844242 = 5.15 \ln_2(32^{32\cdot 32\cdot 32})$	0.299	0.303

Figure 3.1. Extended Cosmos Theory Sequences with sequence F determined by setting λ = 0.303, which is the value of the QED couping constant e which determines the Fine Structure Constant α. The first row is for the case of a multiplier √2. Its n_φ value is an extension of the series of n_φ values 2, 4, 8 of rows A1 through B. The first and second rows are for the cases $n_\varphi = 6$ and $n_\varphi = 8$ which also may be of interest. The second row may correspond to 1.58-bit Large Language Models. Note: $32\cdot 32 = 32^2$ and $32\cdot 32\cdot 32 = 32^3$ above. Note the "Calculated $\lambda_{n\varphi}$" values are determined from the $g_{n\varphi}$ values by $\lambda_{n\varphi} = \pi^2/(\ln g_{n\varphi})^2 + ¼$.

3.1.2 Cosmos Fractal Curve Step Lengths Part

The Cosmos Fractal Curve is similar in form to the Master Equation. It necessarily uses a scaling potential to obtain a well behaved sequence of fractal construction step lengths. This theory only has one spatial dimension q.

\

$$\mathscr{L}_{CC} = (½(\partial A(q)\, \partial A(q)/m - V_{CC}(z)A(q)^2) \tag{3.20}$$

where

$$V_{CC}(q) = \lambda_{CC}/[mq^2] \tag{3.21}$$

Repeating steps similar to the Master Equation case above we find the dynamic equation for the Cosmic Fractal Curve A(q) to be:

$$(\partial^2/\partial q^2 + 1/q\ \partial/\partial q + \lambda_{CC}/mq^2 + mC_{CC})A(q) = 0 \qquad (3.22)$$

with

$$\lambda_{CCC} = \lambda_{CC}/m \qquad (3.23)$$
$$\lambda_{CCC}' = (\lambda_{CCC} - 1/4)^{1/2} \qquad (3.23a)$$

and using the same time as the Master Equation. Substituting in eq. 3.22 we find

$$(\partial^2/\partial q^2 + 1/q\ \partial/\partial q + \lambda_{CCC}/q^2 + mC_{CC})A(q) = 0 \qquad (3.24)$$

where[17] we use the energy $E_{CCn} = (-mC_{CCn})^{1/2}$. Since there is no angular coordinate we obtain the energy (based on Case's energy eigenvalue):

$$E_{CCn} = (-mC_{CCn})^{1/2} = (-mc_{CC})^{1/2}\exp(-n\pi/(\lambda_{CCC} - 1/4)^{1/2}) \qquad (3.25)$$

where c_{CC} is an arbitrary constant. Upon letting[18] n → -n, we have

$$C_{CCn} = c_{CC}\exp(2\pi n/(\lambda_{CCC} - 1/4)^{1/2}) \qquad (3.26)$$

where n is a radial quantum number, and c_{CC} is an arbitrary constant.[19] The growth factor (multiplier) for the Cosmos Fractal Curve piecewise linear curve length steps is

$$g_{CC} = \exp(\pi/(\lambda_{CCC} - 1/4)^{1/2}) \qquad (3.27)$$

Upon letting[20] n → -n, we have

$$E_{CCn} = s_1\exp(n\pi/(\lambda_{CCC} - 1/4)^{1/2}) = s_1 g_{CC}^{\,n} = s_1 2^n \qquad (3.28)$$

using eq. 3.30 below where $s_1 = (-mc_{CC})^{1/2}$ is a constant defined in Fig. 1.1 for the Cosmos Fractal Curve length steps where n is the step number and is now a quantum number.

The growth factor (multiplier) for the Cosmos Fractal Curve step lengths is

$$g_{CC} = 2 \qquad (3.29)$$

in the fractal curve construction if we set

---

[17] The eigenvalue spectrum results from the imposition of orthogonality condition on the solution function near u = 0 according to von Neumann. The value of $-mC_{CC}$ can be taken to be positive since its sign can be changed by Wick rotations of the time variable: t → it.

[18] The sign of n can be positive or negative due to the scaling property of the Master Equation.

[19] The constant c_{CC} can be positive or negative. Its sign can be changed by Wick rotations of the time variable: t → it.

[20] The sign of n can be positive or negative due to the scaling property of the Master Equation.

$$\lambda_{CCC} = n_{CC}^2/m = 20.79 \quad (3.30)$$

based on the form of eq. 3.13 using the values in row A1 of Fig. 3.1. The constant n_{CC} is

$$n_{CC} = 4 \quad (3.31)$$

to good approximation.

Note the A1 row for Cosmos line step lengths where $n_\varphi = 4$ and $g_g = g_{n\varphi=4} = 2$ corresponds to the sequence of steps $s_1 2^n$. Thus the construction steps (first) component of Cosmos Theory is determined.

3.1.3 Cosmos Spaces Part

The Cosmos Spaces has an equation that is similar in form to the Master Equation. It is necessarily a scaling potential to obtain a well behaved power sequence of spaces. This theory has only one spatial dimension p and one field B. The time is the same as the Master Equation time. The Lagrangian with the "CS" label used for Cosmos Spaces is

$$\mathcal{L}_{CS} = = (½\partial B(p)\partial B(p)/m - V_{CS}(z)B(p)^2) \quad (3.32)$$

where

$$V_{CS}(p) = \lambda_{CS}/[m^2p^2] \quad (3.33)$$

Repeating a similar procedure to the Cosmos Fractal Curve construction above we find

$$(\partial^2/\partial p^2 + 1/p\ \partial/\partial p + \lambda_{CS}/mp^2 + mC_{CS})B(p) = 0 \quad (3.34)$$

where the energy

$$E_{CSn} = (-mC_{CSn})^{½} \quad (3.35)$$

We obtain the dynamic equation for the Cosmic Space field B(q) to be:

$$(\partial^2/\partial p^2 + 1/p\ \partial/\partial p + \lambda_{CS}/mp^2 + mC_{CS})B(p) = 0 \quad (3.36)$$

Setting

$$\lambda_{CCS} = \lambda_{CS}/m \quad (3.37)$$
$$\lambda_{CCS}' = (\lambda_{CCS} - ¼)^{½} \quad (3.37a)$$

Substituting in eq. 3.36 we find

$$(\partial^2/\partial p^2 + 1/p\ \partial/\partial p + \lambda_{CCS}/p^2 + mC_{CS})B(p) = 0 \quad (3.38)$$

Since there is no angular coordinate we obtain the energy (based on Case's energy eigenvalue):

$$E_{CSn} = (-mC_{CSn})^{½} = (-mc_{CS})^{½}\exp(-n\pi/(\lambda_{CCS} - ¼)^{½}) \quad (3.39)$$

where c_{CS} is an arbitrary constant. Upon letting[21] n → -n, we have

$$C_{CSn} = c_{CS} \exp(2\pi n/(\lambda_{CCS} - ¼)^{½}) \tag{3.40}$$

where n is a radial quantum number, and c_{CS} is an arbitrary constant.[22] The growth factor (multiplier) for the Cosmos Fractal Curve piecewise linear curve length steps is

$$g_{CS} = \exp(\pi/(\lambda_{CCS} - ¼)^{½}) \tag{3.41}$$

Upon letting[23] n → -n, we have

$$E_{CSn} = s_2 \exp(n\pi/(\lambda_{CCS} - ¼)^{½}) = s_2 g_{CS}{}^n = s_2 2^n \tag{3.42}$$

using eq. 3.43 below where s_2 is a constant[24] defined in Fig. 1.1 for the Cosmos Spaces where n is a quantum number.

The growth factor (multiplier) for the Cosmos Spaces is

$$g_{CS} = 2 \tag{3.43}$$

The g_{CS} value follows if we set

$$\lambda_{CCS} = n_{CS}{}^2/m = 20.79 \tag{3.44}$$

based on the form of eq. 3.13 using the values in row A2 of Fig. 3.1. The constant

$$n_{CS} = 4 \tag{3.45}$$

to good approximation.

Note the A2 row for Cosmos Spaces where $n_\varphi = 4$ and $g_g = g_{n\varphi=4} = 2$ corresponds to the sequence of steps $s_2 2^n$. Thus the construction steps (first) component of Cosmos Theory is determined.

---

[21] The sign of n can be positive or negative due to the scaling property of the Master Equation.
[22] The constant c_{CS} can be positive or negative. Its sign can be changed by Wick rotations of the time variable: t → it.
[23] The sign of n can be positive or negative due to the scaling property of the Master Equation.
[24] The constant s_2 can be positive or negative. Its sign can be changed by Wick rotations of the time variable: t → it.

4. Components of the Components

The three scts of Cosmos equations can be coalesced to form a single dynamics. This dynamics leads to a unified Quantum view of the Cosmos in which all components that determine the *structure* of the Cosmos are combined.

4.1 Unification of the Dynamic Equations' Components of the Cosmos

The unifying differential equation is:

$$\begin{aligned}[\partial^2/\partial u^2 + 1/u\ \partial/\partial u + \partial^2/\partial\varphi^2 + \lambda/mu^2 + mC + \\ + \partial^2/\partial q^2 + 1/q\ \partial/\partial q + \lambda_{CC}/mq^2 + mC_{CC} + \\ + \partial^2/\partial p^2 + 1/p\ \partial/\partial p + \lambda_{CS}/mp^2 + mC_{CS}]\psi = 0\end{aligned} \tag{4.1}$$

with the total energy

$$E^2 = -(mC + mC_{CC} + mC_{CS}) \tag{4.2}$$

This equation in four parameters is separable *in the two dimension Cosmos space generated by the Cosmos Fractal Curve*. The previous subsections calculate the differential equations of the Cosmos components.

The Cosmos wave function has the form:

$$\Psi(u, \varphi, q, p) = \psi_{n\varphi n}(u)A(q)B(p) \tag{4.3}$$

Eq. 3.46 above can be viewed as the consequence of a four dimension space and a time[25] dimension totaling to a five dimension space-time.

4.2 Reconciliation of the Components' Dynamic Equations

We now put the individual components equations and eigenvalues on a common footing.

The energy eigenvalues of the three dynamic equations has the form

$$E_i = s_i 2^{n_i} \tag{4.4}$$

for i = 1, 2, 3.We can now assume that we can adjust the constants s_i and the powers n_i to give all three energies the value

$$E_i = 2^n \tag{4.4}$$

where $n = n_i$ for i = 1, 2, 3. We note the three energies have the form

[25] The time parameter is the distance along the Cosmos Fractal Curve. It arises, as a self-generated, parameter in the three component equations of the Cosmos in chapter 3 simultaneously with the components n_i their equations.

$$E_i = (mC_i)^{1/2} \quad (4.5)$$

And

$$C_i = E_i^2/m \quad (4.6)$$

The solutions of the dynamic equations are

Master Equation Solution

$$\psi_{E3,\, k,\, n_\varphi,\, n}(u, t) = e^{iE_3t} \exp^{ik\varphi} \psi_{3n_\varphi n}(u) \quad (4.7)$$

with the $\psi(u)$ radial factor solution is

$$\psi_{3n_\varphi n}(u) \approx B\cos((\lambda_{n_\varphi} - 1/4)^{1/2} \ln((mC_{3\varphi n})^{1/2} u) + b) \quad (4.8)$$

where b and B are constants. We set B = 1 and b = 0 to obtain (the unnormalized)

$$\begin{aligned} \psi_{E3,\, k,\, n_\varphi,\, n}(u, t) &= e^{iE_3t} \exp^{ik\varphi} \cos((\lambda_{n_\varphi} - 1/4)^{1/2} \ln(2^n u)) \\ &= e^{iE_3t} \exp^{ik\varphi} \cos(\ln(2^n u)^{\lambda'}) \end{aligned} \quad (4.9)$$

where

$$E_3 = 2^n$$
$$\lambda'_{n_\varphi} = \lambda_{n_\varphi} - 1/4)^{1/2}$$

Cosmos Fractal Step Lengths Solution

The field equation for Cosmos step lengths of section 3.1.2 is

$$A_{Ecc,\, n}(q, t) = e^{iE_{cc}t} \cos(\ln(2^n q)^{\lambda_{CCS}'}) \quad (4.10)$$

where

$$\lambda_{CCS}' = (\lambda_{CCS} - 1/4)^{1/2} \quad (4.11)$$

Cosmos Spaces Part Solution

The field equation for Cosmos step lengths of section 3.1.3 is

$$B_{Ecs,\, n}(p, t) = e^{iE_{cs}t} \cos(\ln(2^n p)^{\lambda_{CS}'}) \quad (4.12)$$

where

$$\lambda_{CS}' = (\lambda_{CS} - 1/4)^{1/2} \quad (4.13)$$

In the next chapter we will determine the joint Fourier representation of Ψ in terms of raising and lowering operators, their commutation relations, and the representation of Cosmos structure in terms of Dirac states.

5. Coalescence of the Cosmos Quantum Field Representations

We now consider a joint coalesced formulation of the quantum field for eq. 4.1 that goes beyond the product formulation in eq. 4.2. A product formulation would lead to difficulties with equal time commutation relations.

The coalesced formulation relies on the separation:

$$\cos(\ln(2^n X)^{\lambda_X}) \tag{5.1}$$

for coordinate parameter X and constant λ_X into positive and negative "frequency" parts:

$$\begin{aligned}\cos(\ln(2^n X)^{\lambda_X}) &= (\exp(i\ln(2^n X)^{\lambda_X}) + \exp(-i\ln(2^n X)^{\lambda_X}))/2 \\ &= ((2^n X)^{i\lambda_X} + (2^n X)^{-i\lambda_X})/2\end{aligned} \tag{5.2}$$

Treating the quantum field as spin 0 we find the general form *that satisfies eq. 4.1*:

$$\begin{aligned}\psi_{n_\varphi n_1,\, n_2, n_3}(\varphi, u, q, p, t) - &\int dk\, N(k,n_1,n_2,n_3)\{a(k,n_1,n_2,n_3)\, e^{-ik\varphi}\exp(-i2^{n_1+n_2+n_3}t)e^{-ik\varphi}\,\cdot \\ &\cdot (2^{n_3}u)^{-i\lambda'_{n_\varphi}}(2^{n_2}p)^{-i\lambda_{CCC}'}(2^{n_1}q)^{-i\lambda_{CCS}'} + \\ &+ a^\dagger(k,n_1,n_2,n_3)\exp(i2^{n_1+n_2+n_3}t)e^{ik\varphi}(2^{n_3}u)^{i\lambda'_{n_\varphi}}(2^{n_1}q)^{i\lambda_{CCC}'}(2^{n_2}p)^{-i\lambda_{CCS}'}\}\end{aligned} \tag{5.3}$$

where the n_1, n_2, n_3 quantum numbers represent the Cosmos Fractal step numbers, the Cosmos field numbers and the Master Equation solution numbers n_{CC}, n_{CS}, n_{n_φ} respectively, using the λ values:

$$\lambda_{n_\varphi}' = (\lambda_{n_\varphi} - 1/4)^{1/2} \tag{3.13b}$$
$$\lambda_{CCC}' = (\lambda_{CCC} - 1/4)^{1/2} \tag{3.23a}$$
$$\lambda_{CCS}' = (\lambda_{CCS} - 1/4)^{1/2} \tag{3.37a}$$

Summing over the n_i values we find

$$\begin{aligned}\psi_{n_\varphi}(\varphi, u, q, p, t) = \sum_{n_1, n_2, n_3} &\int dk\, N(k,n_1,n_2,n_3)\{a(k,n_1,n_2,n_3)\, e^{-ik\varphi}\exp(-i2^{n_1+n_2+n_3}t)e^{-ik\varphi}\,\cdot \\ &\cdot (2^{n_3}u)^{-i\lambda'_{n_\varphi}}(2^{n_2}p)^{-i\lambda_{CCC}'}(2^{n_1}q)^{-i\lambda_{CCS}'} + \\ &+ a^\dagger(k,n_1,n_2,n_3)\exp(i2^{n_1+n_2+n_3}t)e^{ik\varphi}(2^{n_3}u)^{i\lambda'_{n_\varphi}}(2^{n_1}q)^{i\lambda_{CCC}'}(2^{n_2}p)^{-i\lambda_{CCS}'}\}\end{aligned} \tag{5.3a}$$

5.1 Total Coalesced Field ψ

We can now combine the three components into a totally coalesced field. We set the λ's to be equal since they can set $n_{n_\varphi} = n_{CC} = n_{CS} = n = 4$ as in chapters 3 and 4:

$$\lambda = \lambda_{n_\varphi}' = \lambda_{CCC}' = \lambda_{CCS}' = \lambda_{CCS}' = 4 \tag{5.4}$$

Then eq. 5.3a becomes

$$\psi_{n\varphi}(\varphi, u, q, p, t) = \sum_n \int dk\ N(k, n)\{a_{n\varphi}(k, n)\exp(-i2^{3n}t)e^{-ik\varphi}(2^{3n}uqp)^{-i\lambda} +$$

$$+ a_{n\varphi}^{\dagger}(k, n)\exp(i2^{3n}t)e^{ik\varphi}(2^{3n}uqp)^{i\lambda}\} \tag{5.5}$$

where n replaces the common value of the quantum numbers. Noting the coordinates are only in the form

$$x = uqp \tag{5.5a}$$

we change eq. 5.5 to

$$\psi_{n\varphi}(\varphi, x, t) = \sum_n \int dk\ N(k, n)\{a_{n\varphi}(k, n)\exp(-i2^{3n}t)e^{-ik\varphi}(2^{3n}x)^{-i\lambda} +$$

$$+ a_{n\varphi}^{\dagger}(k, n)\exp(i2^{3n}t)e^{ik\varphi}(2^{3n}x)^{i\lambda}\} \tag{5.6}$$

We extend the definition of the quantum number n, which also relates to the space-time dimension of Cosmos spaces, by making it continuous with the result:

$$\psi_{n\varphi}(\varphi, x, t) = \int dndk\ N(k, n)\{a_{n\varphi}(k, n)\exp(-i2^{3n}t)e^{-ik\varphi}(2^{3n}x)^{-i\lambda} +$$

$$+ a_{n\varphi}^{\dagger}(k, n)\exp(i2^{3n}t)e^{ik\varphi}(2^{3n}x)^{i\lambda}\} \tag{5.7}$$

We now introduce a continuous valued momentum and energy based on the form of ψ:

$$\begin{aligned} p &= 2^{3n} \\ E &= p \end{aligned} \tag{5.8}$$

treating 2^{3n} as a momentum. One can view the field as representing a massless boson. The relation of p and n gives

$$dn/dp = 1/3p$$

implying

$$\psi_{n\varphi}(\varphi, x, t) = 1/3 \int d(\ln p)dk\ N(k, p)\{a_{n\varphi}(k, p)\exp(-iEt)e^{-ik\varphi}(px)^{-i\lambda} +$$

$$+ a_{n\varphi}^{\dagger}(k, p)\exp(iEt)e^{ik\varphi}(px)^{i\lambda}\} \tag{5.9}$$

$$\psi_{n\varphi}(\varphi, X, t) = 1/3 \int dPdk\ N(k, P)\{a_{n\varphi}(k, P)e^{-iEt - ik\varphi - i\lambda X - i\lambda P} + a_{n\varphi}^{\dagger}(k, P)e^{iEt + ik\varphi + i\lambda X + i\lambda P}\} \tag{5.10}$$

where

$$\begin{aligned} X &= \ln x \\ P &= \ln p \end{aligned} \tag{5.11}$$

We now define the commutation relations

$$[a_{n_\varphi}(k, P), a_{n_\varphi'}^{\dagger}(k, P)] = \delta_{n_\varphi,n_\varphi'}\delta(k - k')\,\delta(P - P') \quad (5.12)$$

We now set

$$[\psi_{n_\varphi}(\varphi, X, t), \psi_{n_\varphi'}'(\varphi', X', t)] = i\,\delta_{n_\varphi,n_\varphi'}\delta(\varphi - \varphi')\delta(X - X') \quad (5.13)$$

where

$$\psi_{n_\varphi'}'(\varphi', X', t) = \partial\psi_{n_\varphi'}(\varphi', X', t)/\,\partial t \quad (5.13a)$$

to determine the normalization factor N(k, P).

Note:

$$X = \ln x = \ln(uqp) \quad (5.14)$$
$$P = \ln p = \ln(2^{3n}) = 3n$$
$$E = e^{P}$$

We see the momentum P may be interpreted as indicating a fractal. The three factors, uqp, suggest they may also be fractal with total dimension 3 just as the time t has a fractal origin.

We can also see a possible commutation relation between a Cosmos two dimension space coordinate X and dimension n:

$$[X, n] = i\hbar \quad (5.14a)$$

making the choice of a Cosmos space, construction step, Master Equation number and Cosmos field coordinates "uncertain": An Uncertainty Condition.

5.2 Total Coalesced Field States

The coalesced field has a corresponding set of states. The single "particle" state is

$$|\,n_\varphi, k, P\rangle = a_{n_\varphi}^{\dagger}(k, P)|0\rangle = \int dX d\varphi\, e^{-iEt - ik\varphi - i\lambda X - i\lambda P}\,\psi_{n_\varphi}(\varphi, X, t)\,|0\rangle \quad (5.15)$$

with

$$\langle n_\varphi, k, P|\, n_\varphi', k', P'\rangle = \delta_{n_\varphi,n_\varphi'}\delta(k - k')\,\delta(P - P') \quad (5.15a)$$

We can define a Gaussian superposition of this state with

$$|\,n_\varphi, n, G\rangle = \int dPdk\, \exp(-(P - 3n)^2)/n^2)\,|\,n_\varphi, k, P\rangle \quad (5.16)$$

where the Gaussian is centered on a particular value n. The state $|\,n_\varphi, n, G\rangle$ specifies a superposition of $|\,n_\varphi, k, P\rangle$ states for the nth level. For the fractal curve it is the nth construction step. For the Cosmos spaces component it is the nth space. For the Master Equation it is the nth solution.

We can also define a full state for all n of the components with

$$|\,n_\varphi, G\rangle = \prod_{n=1}^{\infty} |\,n_\varphi, n, G\rangle = \int dPdk\{\prod_{n=1}^{\infty} \exp(-(P - 3n)^2)/n^2)\}|\,n_\varphi, k, P\rangle \quad (5.17)$$

We integrate over P and k to obtain a state summed over these degrees of freedom. If we specify an equal deviation for all n:

$$\varepsilon = P - 3n$$

then we obtain the combined state

$$| n_\varphi, G > = \prod_{n=1}^{\infty} | n_\varphi, n, G > = \int dPdk \exp\{-\sum_{n=1}^{\infty} \varepsilon^2/n^2)\}| n_\varphi, k, P> \tag{5.18}$$

$$= \int dPdk \exp(-\varepsilon^2\pi^2/6) \mid n_\varphi, k, P>$$
$$= \int dPdk \exp(-\varepsilon^2 \zeta(2)) \mid n_\varphi, k, P> \tag{5.19}$$

using the Riemann zeta function. The state $| n_\varphi, G >$ specifies a superposition of $| n_\varphi, k, P>$ states for all n levels. For the fractal curve it is the sum of the construction steps. For the Cosmos spaces component it is the sum of the Cosmos spaces. For the Master Equation it is the sum of the solutions.

To calculate the probability we square the probability amplitude P_A:

$$P_A = < n_\varphi', G' \mid n_\varphi, G > = \int dP'dk' \exp(-\varepsilon^2\pi^2/6) \int dPdk \exp(-\varepsilon^2\pi^2/6)\, \delta_{n_\varphi', n_\varphi} \delta(P - P')$$
$$= \exp(-\varepsilon^2\pi^2/3) \tag{5.20}$$

times a removable infinite factor. The probability is then

$$P = P_A{}^*P_A = N^2 \exp(-2\varepsilon^2\pi^2/3) \tag{5.21}$$

We normalize the probability with

$$N^{-2} = \int_0^\infty d\varepsilon \exp(-2\varepsilon^2\pi^2/3) = 1.225 \tag{5.22}$$

or

$$N = 0.905 \tag{5.23}$$

independent of n_φ.

The probability from eq. 5.21 for the "classical" value for each of the three components where $\varepsilon = 0$ is

$$P = N^2 = 0.82 \tag{5.24}$$

This probability for the "classical" result, obtained in a modest Gaussian model, is encouragingly large.

The probability is for this totally coalesced case where all the components deviate by ε for all n from the values of the "Classical" Cosmos Theory. For the fractal curve component it is the sum of the construction steps. For the Cosmos spaces component it is the sum of the Cosmos spaces. For the Master Equation it is the sum of the solutions and eigenvalues.

5.3 Timeless Form of the Totally Coalesced Field and States

We can put eq. 5.10 in a form that makes the time dependence t, which is measured as the length of the Cosmos Fractal Curve, into a dependence on the construction steps of the Curve specified in section 2.1. The Cosmos two dimension space time is then infinite from $-\infty$ to $+\infty$. Since there is no time outside of universes the Cosmos structure is timeless. We then may say Cosmos structure is eternal.

We can exhibit the infinite dependence of the quantum field on t by defining t in terms of a step number N:

$$t = 2^{N_t} \tag{5.25}$$

resulting in

$$\psi_{n\varphi}(\varphi, X, N_t) = 1/3 \int dPdk\ N(k, P)\{a_{n\varphi}(k, P)e^{-iEt - ik\varphi - i\lambda X - i\lambda P} + a_{n\varphi}^{\dagger}(k, P)e^{iEt + ik\varphi + i\lambda X + i\lambda P}\} \tag{5.26}$$

We may make N_t continuous to obtain a continuous "time."

The individual elements of the field equation for construction step n are specified with:

$$\psi_{n\varphi n}(\varphi, X, N_t) = 1/3 \int dk\ N(k, 3n)\{a_{n\varphi}(k, 3n)e^{-iEt - ik\varphi - i\lambda X - i3\lambda n} + a_{n\varphi}^{\dagger}(k, 3n)e^{iEt + ik\varphi + i\lambda X + i3\lambda n}\} \tag{5.27}$$

The Cosmos structure of four components is timeless and eternal.

5.3.1 Why the Existence of Time Necessitates a Universe

The definition of time and spatial coordinates is necessarily based on the existence of tools that can measure their values from point to point. This fact is particularly evident when one imposes quantization to obtain Uncertainty Relations. Since no tools exist outside of universes one cannot define coordinates in any Physically meaningful way. Thus

There is no time outside of universes.
There are no coordinates outside of universes.

In the case of Cosmos structure there is an entity, the Cosmos Fractal Curve, which exists eternally in our view. We define a space based on its space-filling properties. We define a time in terms of points along the length of the Curve (as if it were straightened). Thus we obtain the coordinates of the two (plus one) dimension space generated from the Curve that is used to determine Cosmos structure. The components of the Cosmos must be thought of as instantaneous, but in the absence of universes (except the Parent universe, which is a non-participant, discussed later), the instantaneity is not at any point in time.

It is an eternal instantaneity.

5.4 Partly Coalesced Field

This case describes partial coalescence where the components are combined in a field but retain their identities. Our starting point is eq. 5.3.

We change the quantum numbers n_1, n_2, and n_3 to continuous parameters. First, we introduce continuous valued momentums and energies:

$$\begin{aligned} p_i &= 2^{n_i} \\ E_i &= p_i \end{aligned} \tag{5.28}$$

treating 2^{n_i} as a momentum. One can view the field as representing a massless boson since $E_i = p_i$.

We also define:

$$\begin{aligned} X_i &= \ln x_i \\ P_i &= \ln p_i = \ln(2^{n_i}) = n_i \end{aligned} \tag{5.29}$$

We see the momentum P_i may be interpreted as indicating a fractal. The three factors X_i suggest they may also be fractal with total dimension 3 just as the time t has a fractal origin.

We can also see a possible commutation relation between a Cosmos Space two dimension space coordinate X_i and dimension n_i:

$$[X_i, n_j] = i\hbar\delta_{ij} \tag{5.30}$$

for i, j = 1, 2, 3 making the choice of a Cosmos dimension and the corresponding Cosmos coordinate "uncertain."

The relation of p_i and n_i gives

$$dn_i/dp_i = 1/p_i \tag{5.31}$$

implying

$$\begin{aligned} \psi_{n_\varphi n_1, n_2, n_3}(\varphi, u, q, p, t) = \int dP_1 dP_2 dP_3 dk\, N(k, P_1,P_2,P_3) \cdot \\ \cdot\{a(k, P_1,P_2,P_3)\, e^{-ik\varphi} \exp(-i2^{n_1+n_2+n_3}t)\, e^{-ik\varphi} (p_3u)^{-i\lambda'_{n_\varphi}} (p_1q)^{-i\lambda_{ccc}'} (p_2p)^{-i\lambda_{ccs}'} + \\ + a^\dagger(k, P_1,P_2,P_3) \exp(i2^{n_1+n_2+n_3}t) e^{ik\varphi} (p_3u)^{i\lambda'_{n_\varphi}} (p_1q)^{i\lambda_{ccc}'} (p_2p)^{-i\lambda_{ccs}'}\} \end{aligned} \tag{5.32}$$

Using

$$(px)^{i\lambda} = \exp(i\lambda(P + X)) \tag{5.33}$$

We obtain

$$\begin{aligned} \psi_{n_\varphi n_1, n_2, n_3}(\varphi, X_1,X_2,X_3, t) = \int dP_1 dP_2 dP_3 dk\, N(k,P_1,P_2,P_3) \cdot \\ \cdot\{a(k,P_1,P_2,P_3)\, e^{-ik\varphi} e^{-i(E_1+E_2+E_3)t} e^{-ik\varphi}\, e^{-i\lambda_{n_\varphi}'(P_3 + X_3)}\, e^{-i\lambda_{ccc}'(P_1 + X_1)}\, e^{-i\lambda_{ccs}'(P_2 + X_2)} + \\ + a^\dagger(k, P_1,P_2,P_3) e^{ik\varphi} e^{i(E_1+E_2+E_3)t}\, e^{i\lambda_{n_\varphi}'(P_3 + X_3)}\, e^{i\lambda_{ccc}'(P_1 + X_1)}\, e^{i\lambda_{ccs}'(P_2 + X_2)}\} \end{aligned}$$

$$\psi_{n\varphi n_1,\, n_2, n_3}(\varphi,\, X_1,X_2,X_3,\, t) = \int dP_1 dP_2 dP_3 dk\, N(k,P_1,P_2,P_3) \cdot$$
$$\cdot\{a(k,P_1,P_2,P_3)\, e^{-ik\varphi} e^{-i(E_1+E_2+E_3)t} e^{-ik\varphi}\, e^{-i\lambda_{n_\varphi}'P_3 - i\lambda_{ccc}'P_1 - i\lambda_{ccs}'P_2}\, e^{-i\lambda_{n_\varphi}'X_3 - i\lambda_{ccc}'X_1 - i\lambda_{ccs}'X_2} +$$
$$+ a^\dagger(k,\, P_1,P_2,P_3) e^{ik\varphi} e^{i(E_1+E_2+E_3)t}\, e^{i\lambda_{n_\varphi}'P_3 + i\lambda_{ccc}'P_1 + i\lambda_{ccs}'P_2}\, e^{i\lambda_{n_\varphi}'X_3 + i\lambda_{ccc}'X_1 + i\lambda_{ccs}'X}\} \qquad (5.34)$$

using eq. 5.29 where the n_1, n_2, n_3 quantum numbers represent the Cosmos Fractal step numbers, the Cosmos spaces' field numbers, and the Master Equation solution numbers n_{CC}, n_{CS}, n_{n_φ} respectively, using the λ values:

$$\lambda_{n_\varphi}' = (\lambda_{n_\varphi} - 1/4)^{1/2} = \lambda_3' \qquad (5.35)$$
$$\lambda_{CCC}' = (\lambda_{CCC} - 1/4)^{1/2} = \lambda_1' \qquad (5.36)$$
$$\lambda_{CCS}' = (\lambda_{CCS} - 1/4)^{1/2} = \lambda_2' \qquad (5.37)$$

and

$$E_i = e^{P_i} \qquad (5.38)$$

For i = 1, 2, 3. We now define the commutation relations

$$[a_{n_\varphi}(k,P_1,P_2,P_3),\, a_{n_\varphi'}^\dagger(k',P_1',P_2',P_3')] = \delta_{n\varphi,n\varphi'}\delta(k - k')\, \delta(P_1 - P_1')\, \delta(P_2 - P_2')\, \delta(P_3 - P_3') \qquad (5.39)$$

We set the equal time commutation relations:

$$[\psi_{\,n\varphi n_1,\, n_2,n_3}(\varphi,\, X_1,X_2,X_3,t),\, \psi_{n\varphi n_1,\, n_2,n_3}'(\varphi',\, X_1',X_2',X_3',\, t)] =$$
$$= i\, \delta_{n\varphi,n\varphi'}\delta(\varphi - \varphi')\, \delta(X_1 - X_1')\, \delta(X_2 - X_2')\, \delta(X_3 - X_3') \qquad (5.40)$$

where

$$\psi_{\,n\varphi n_1,\, n_2,n_3}'(\varphi,\, X_1,X_2,X_3,\, t) = \partial_{n\varphi n_1,\, n_2,n_3}\, \psi(\varphi,\, X_1,X_2,X_3,\, t)/\partial t \qquad (5.41)$$

to determine the normalization factor $N(k,\, P_1,P_2,P_3)$.

5.5 Partly Coalesced Field States

The coalesced field has a corresponding set of states. The single "particle" state is

$$|\, n\varphi,\, k,\, P_1,P_2,P_3\rangle = a_{n_\varphi}^\dagger(k,\, P_1,P_2,P_3)|0\rangle =$$
$$= \int dX_1 dX_2 dX_3 d\varphi e^{-iEt - ik\varphi}\, e^{-i\lambda_{n_\varphi}'P_3 - i\lambda_{ccc}'P_1 - i\lambda_{ccs}'P_2}\, e^{-i\lambda_{n_\varphi}'X_3 - i\lambda_{ccc}'X_1 - i\lambda_{ccs}'X_2} \psi_{n\varphi}(\varphi,\, X_1,X_2,X_3,\, t)\, |0\rangle \qquad (5.42)$$

The orthogonality condition is

$$\langle n\varphi,\, k,\, P_1,P_2,P_3|\, n\varphi',\, k',\, P_1',P_2',P_3'\rangle = \delta_{n\varphi,n\varphi'}\delta(k - k')\, \delta(P_1 - P_1')\, \delta(P_2 - P_2')\delta(P_3 - P_3') \qquad (5.43)$$

We can form states which are superpositions of HyperCosmos and 2nd Kind HyperCosmos spaces such as

$$|\, n\varphi,\, k,\, P_1,\, P_3,\, \theta\rangle = \cos(\theta)\, |\, n\varphi,\, k,\, P_1,\, r+4,\, P_3\rangle + \sin(\theta)\, |\, n\varphi,\, k,\, P_1,\, r+3,\, P_3\rangle \qquad (5.44)$$

where r is the Cosmos spaces dimension and $P_2 = r + 4$ specifies a HyperCosmos space and $P_2 = r + 3$ specifies a 2nd Kind HyperCosmos space. The angle θ specifies the mix of spaces. We describe the Physics of the mixed state in detail in Blaha (2025h).

We can create a mixed HyperCosmos–2nd Kind HyperCosmos spaces contracted state from eq. 5.44 using a variation of the PseudoFermion formalism of Blaha (2024j):

$$| n_\varphi, P_2> = \int dP_1 dP_3 dk\, | n_\varphi = 4, k, P_1, P_2, P_3, \theta> \tag{5.45}$$

The mixed state is

$$\begin{aligned} | n_\varphi, \theta> &= \int dP_1 dP_3 dk\, [\cos(\theta)\, | n_\varphi, k, P_1, r+4, P_3> + \sin(\theta)\, | n_\varphi, k, P_1, r+3, P_3>] \\ &= \cos(\theta)\, | n_\varphi, P_2 = r+4> + \sin(\theta)\, | n_\varphi, k, P_2 = r+3> \end{aligned} \tag{5.45a}$$

We integrate over P_1, P_3, and k to obtain a state summed over these degrees of freedom. This state is considered in chapter 6, and in sections 2.1.3 and 2.1.4 of Blaha (2025h). There we give an interpretation of mixed HyperCosmos–2nd Kind HyperCosmos spaces' states.

We can also define a Gaussian superposition of quantum Cosmos Fractal Curve construction steps summed over the other Cosmos components. The quantum nth step is:

$$| n_\varphi, n, G> = \int dk dP_1 dP_2 dP_3 \exp(-(P_1 - n)^2)/n^2)\, | n_\varphi, k, P_1, P_2, P_3> \tag{5.46}$$

where the Gaussian is centered on a particular value n. The state $| n_\varphi, n, G >$ specifies a superposition of $| n_\varphi, k, P_1, P_2, P_3>$ states for the nth level. For the fractal curve it is the nth construction step.

We can also define a full state for all n of the components with

$$| n_\varphi, G > = \prod_{n=1}^{\infty} | n_\varphi, n, G > = \int dP_1 dP_2 dP_3 dk \{ \prod_{n=1}^{\infty} \exp(-(P_1 - 3n)^2)/n^2) \}| n_\varphi, k, P_1, P_2, P_3> \tag{5.47}$$

If we specify an equal deviation for all n:

$$\varepsilon = P_1 - n \tag{5.48}$$

then we obtain the combined state

$$\begin{aligned} | n_\varphi, G > &= \prod_{n=1}^{\infty} | n_\varphi, n, G > = \int dP_1 dP_2 dP_3 dk \exp\{-\sum_{n=1}^{\infty} \varepsilon^2/n^2)\}| n_\varphi, k, P_1, P_2, P_3> \\ &= \int dP_1 dP_2 dP_3 dk \exp(-\varepsilon^2\pi^2/6)\, | n_\varphi, k, P_1, P_2, P_3> \end{aligned} \tag{5.49}$$

The state $| n_\varphi, G >$ specifies a superposition of $| n_\varphi, k, P_1, P_2, P_3>$ states for all n levels. For the fractal curve it is the sum of the construction steps. To calculate the probability we square the probability amplitude P_A:

$$P_A = \langle n_\varphi', G' \mid n_\varphi, G \rangle = \int dP'dk' \exp(-\varepsilon^2\pi^2/6) \int dPdk \exp(-\varepsilon^2\pi^2/6)\, \delta_{n_\varphi', n_\varphi} \delta(P - P')$$
$$= \exp(-\varepsilon^2\pi^2/3) \qquad (5.50)$$

times a removable infinite factor. The probability is then

$$P = P_A{}^*P_A = N^2 \exp(-2\varepsilon^2\pi^2/3) \qquad (5.51)$$

We normalize the probability with

$$N^{-2} = \int_0^\infty d\varepsilon \exp(-2\varepsilon^2\pi^2/3) = 1.225 \qquad (5.52)$$

or

$$N = 0.905 \qquad (5.53)$$

independent of n_φ.

The probability from eq. 5.51 for the "classical" value of the complete set of construction steps where $\varepsilon = 0$ is

$$P = N^2 = 0.82 \qquad (5.54)$$

This probability for the "classical" result, obtained in this modest Gaussian model, is encouragingly large.

This probability is for the partly coalesced case where all the construction steps deviate by ε for all n from the values of the "Classical" Cosmos Theory. These results appear in chapter 3 of Blaha (2025h).

6. Extension to Include the Parent Universe

Cosmos Theory posits a Parent universe that is coeval with its four components. The Parent universe is ever present just as the four components. We previously showed that a Consistency Condition exists that provides a limit on the number of Cosmos spaces and on the size of the largest space. In this chapter we develop the Quantum Cosmos theory of a Parent universe including a Quantum formulation of the Consistency Condition, and a formulation of Quantum subuniverses (child universes).

6.1 Cosmos Space – Parent Universe Quantum Field in r Dimensions

Treating the Parent universe as a spin 0 quantum field we consider the generalization of eqs. 3.1, 4.1 and 5.3 to include the Parent(s) universe Lagrangian:

$$\mathcal{L} = \mathcal{L}_{CC} + \mathcal{L}_{CS} + \mathcal{L}_{ME} + \mathcal{L}_{Parent} \tag{6.1}$$

The dynamic equations are

$$\begin{aligned}[\partial/\partial X_i \partial/\partial X^i + M^2 &+ \partial^2/\partial u^2 + 1/u\ \partial/\partial u + \partial^2/\partial\varphi^2 + \lambda/mu^2 + mC + \\ &+ \partial^2/\partial q^2 + 1/q\ \partial/\partial q + \lambda_{CC}/mq^2 + mC_{CC} + \\ &+ \partial^2/\partial p^2 + 1/p\ \partial/\partial p + \lambda_{CS}/mp^2 + mC_{CS}]\psi(X, u, q, p) = 0\end{aligned} \tag{6.2}$$

where the X_i are the coordinates in the Cosmos two dimension space and the mass-energy is M after separating the Cosmos space time t dependence. (Time is measured along the length of the Cosmos Fractal Curve.) The Parent universe has the spatial coordinates X_i within the two dimension Cosmos Space. Note the u, q and p coordinates are also within the two dimension Cosmos Space. Again the absence of a true time within a universe makes the dynamics instantaneously effected.

Following the derivation of the quantum field's u, p, and q dependence, and introducing X dependence, we obtain:

$$\begin{aligned}\psi_{n_\varphi n_1, n_2, n_3}(\varphi, u,q,p,X,t) &= \sum_{n_1, n_2, n_3} \int d^2s\,dk\,N(k,n_1,n_2,n_3,s)\{a(k,n_1,n_2,n_3,s)e^{-ik\varphi}\exp(-i2^{n_1+n_2+n_3}t)e^{-ik\varphi} \cdot \\ &\cdot e^{-is\cdot X}(2^{n_3}u)^{-i\lambda'_{n_\varphi}}(2^{n_2}p)^{-i\lambda_{ccc}'}(2^{n_1}q)^{-i\lambda_{ccs}'} + \\ &+ a^\dagger(k,n_1,n_2,n_3,s)e^{is\cdot X}\exp(i2^{n_1+n_2+n_3}t)e^{ik\varphi}(2^{n_3}u)^{i\lambda'_{n_\varphi}}(2^{n_1}q)^{i\lambda_{ccc}'}(2^{n_2}p)^{-i\lambda_{ccs}'}\}\end{aligned} \tag{6.3}$$

where the n_1, n_2, n_3 quantum numbers represent the Cosmos Fractal step numbers, the Cosmos field numbers and the Master Equation solution numbers n_{CC}, n_{CS}, n_{n_φ} respectively, using the λ values:

$$\lambda_{n_\varphi}' = (\lambda_{n_\varphi} - 1/4)^{1/2} \tag{3.13b}$$

$$\lambda_{CCC}' = (\lambda_{CCC} - 1/4)^{1/2} \tag{3.23a}$$

$$\lambda_{CCS}' = (\lambda_{CCS} - 1/4)^{1/2} \tag{3.37a}$$

We now define the commutation relations (Note $P_i = n_i$) for i = 1, 2, 3.)

$$[a_{n_\varphi}(k,P_1,P_2,P_3,s),\ a_{n_\varphi'}^{\dagger}(k',P_1',P_2',P_3',s')] = \\ = \delta_{n_\varphi,n_\varphi'}\delta(k - k')\ \delta^2(s - s')\delta(P_1 - P_1')\ \delta(P_2 - P_2')\ \delta(P_3 - P_3') \tag{6.4}$$

where s is the momentum conjugate to X. The equal time commutation relations are:

$$[\psi_{n_\varphi n_1,\ n_2,n_3}(\varphi,\ x_1,x_2,x_3,X,t),\ \psi'_{n_\varphi n_1,\ n_2,n_3}(\varphi',\ x_1',x_2',x_3',X',t)] = \\ = i\ \delta_{n_\varphi,n_\varphi'}\delta(\varphi - \varphi')\ \delta(x_1 - x_1')\ \delta(x_2 - x_2')\ \delta(x_3 - x_3')\ \delta^2(X - X') \tag{6.5}$$

where $x_1 = u$, $x_2 = q$, $x_3 = p$ and

$$\psi'_{n_\varphi n_1,\ n_2,n_3}(\varphi,\ x_1,x_2,x_3,X,t) = \partial_{n_\varphi n_1,\ n_2,n_3}\ \psi(\varphi,\ x_1,x_2,x_3,X,t)/\partial t \tag{6.6}$$

to determine the normalization factor $N(k,\ P_1,P_2,P_3,s)$.

6.2 Parent Universe

The coalesced field has a corresponding set of states. The single "particle" state is

$$|\ n_\varphi,\ k,\ P_1,P_2,P_3,s > = a_{n_\varphi}^{\dagger}(k,\ P_1,P_2,P_3,s)|0> = \\ = \int d^2X dx_1 dx_2 dx_3 d\varphi e^{-iEt-ik\varphi}\ e^{-is\cdot X} e^{-i\lambda_{n_\varphi}'P_3 - i\lambda_{ccc}'P_1 - i\lambda_{ccs}'P_2}\ e^{-i\lambda_{n_\varphi}'x_3 - i\lambda_{ccc}'x_1 - i\lambda_{ccs}'x_2}\ \cdot \\ \cdot\ \psi_{n_\varphi}(\varphi,\ x_1,x_2,x_3,X,\ t)\ |0> \tag{6.7}$$

We can form states which are superpositions of HyperCosmos and 2nd Kind HyperCosmos spaces such as

$$|\ n_\varphi,\ k,\ P_1,\ P_3,\ s,\ \theta> = \cos(\theta)\ |\ n_\varphi,\ k,\ P_1,\ r+4,\ P_3,\ s> + \sin(\theta)\ |\ n_\varphi,\ k,\ P_1,\ r+3,\ P_3,\ s> \tag{6.8}$$

where r is the Cosmos spaces dimension and $P_2 = r + 4$ specifies a HyperCosmos space and $P_2 = r + 3$ specifies a 2nd Kind HyperCosmos space. The angle θ specifies the mix of spaces. We describe the Physics of the mixed state in detail in Blaha (2025h).

6.2.1 Mixed Parent Universe Field

We can create a mixed HyperCosmos–2nd Kind HyperCosmos Parent space contracted state from eq. 6.8 using a variation of the PseudoFermion formalism of Blaha (2024j):

$$|\ n_\varphi,\ P_2,\ s> = \int dP_1 dP_3 dk\ |\ n_\varphi = 4,\ k,\ P_1,\ P_2,\ P_3,\ s,\ \theta> \tag{6.9}$$

The mixed state is

$$|n_\varphi,\ s,\ \theta> = \int dP_1 dP_3 dk\ [\cos(\theta)|\ n_\varphi,k,P_1,P_2 = r+4,P_3,s> + \sin(\theta)|\ n_\varphi,k,P_1,P_2 = r+3,P_3,s>] \\ = [\cos(\theta)\ |\ n_\varphi,\ P_2 = r+4,\ s> + \sin(\theta)\ |\ n_\varphi,\ k,\ P_2 = r+3,\ s>] \tag{6.10}$$

We integrate over P_1, P_3, and k to obtain a state summed over these degrees of freedom. This state is considered in chapter 7 and sections 2.1.3 and 2.1.4 of Blaha (2025h). These sections give an interpretation of mixed HyperCosmos–2nd Kind HyperCosmos spaces' states.

6.3 Quantum Consistency Condition for the Number of HyperCosmos Spaces and for a Parent Universe[26]

There is something of a quandary in the definition of a structure for Cosmos spaces and the separate definition/origin of universes. Previously, we have pointed out the importance of a separation of structure (the Cosmos spaces[27] spectrum) from substance (universes).

But the question of their timing (for example, which appeared first?) arises. A space first? Or a time first? Since time is part of a coordinate system, which implicitly or explicitly requires a universe, we found there must be a preexisting universe for time to be defined. However, a universe must be defined based on the definition of a space. Thus we can only conclude the initial *Physical* Cosmos Space and Parent universe must be simultaneous in the sense that they cannot be distinguished by a time interval.

Since there is no time before the existence of the Parent universe we cannot specify a creation time. The Parent universe, and Cosmos Space, must thus be viewed as without a beginning. Should the Parent universe cease to exist then time ceases as well. Thus we have an eternal initial universe that we call the *Parent*[28] universe. Within it, "subuniverses" may be created according to the time of the Parent universe.

The dimension array of the Cosmos Space and its corresponding Parent universe must be simultaneous.

6.4 Quantum Consistency Condition

The set of possible HyperCosmos spaces initially appears unlimited. Its dimension r starts at r = 0. It has no apparent upper bound on dimension. We seek a *Physical* basis for an upper bound on space dimension to obtain a finite set of Cosmos spaces – a convenient and expedient requirement.

We have suggested a consistency condition for the maximum Cosmos space and Parent universe dimension. The *Physical* principle, which we specified, is to require the Parent universe to not collapse but to be of fixed size or expanding so that it persists eternally. To implement this principle we require a consistency condition that prevents the collapse of the Parent universe. We therefore require the Parent universe be the universe of least dimension (greater than zero) where the virtual thermodynamic outward pressure within the Parent universe is greater than, or equal to, the presumptive

[26] See Blaha (2025c), *Fundamental Constants of Cosmos Theory and The Standard Model.* Blaha (2024f) and also Blaha (2022d) and (2023e).

[27] The Cosmos spaces spectrum is defined from the consideration of totally antisymmetric tensors and the creation and annihilation operators of a fermion operator Fourier expansion. It has an equivalent view as a fractal curve that we have constructed and called the Cosmos Fractal Curve.

[28] There could be several "simultaneous" Parent universes.

external Casimir vacuum inward pressure at the time of origin of the Parent universe.[29] See Fig. 6.1 below. Below this minimum dimension the Parent universe cannot exist since the virtual vacuum pressure would cause it to be contracted into non-existence.

We can summarize the consistency condition as:[30]

The Parent universe is of the least dimension such that it doesn't contract.

The dimension r of the Parent universe then fixes a maximum *Physical* Cosmos space dimension.[31] A Parent universe of lesser dimension would collapse. A Parent universe of greater dimension may be stable or collapse.

Within the Parent universe there are subuniverses of lower dimension corresponding to lower dimension HyperCosmos spaces. These universes are not subject to this consistency condition and may expand, be stable or collapse.

Recently we suggested a Consistency Condition that specifies the number of HyperCosmos spaces as 10 and also the existence of an 18 dimension Parent universe.

The basis of this analysis was the existence of a Cosmos spaces structure determined from an analysis of totally asymmetric tensors and the creation/annihilation operators of a fermion field.

We now restate the analysis that leads to the specification of the size of the set of *Physical* Cosmos spaces as numbering 10, and to the dimension of the Parent space to be 18.[32]

6.5 Balance Between Expansion and Contraction

When a possible virtual universe is "virtually" created it may expand (Hubble expansion), or it may contract to a point and then cease to exist. The decision is based on the dynamics of the virtual creation. Does the virtual universe possess sufficient mass-energy to generate a pressure for expansion or does the surrounding[33] virtual vacuum pressure cause the universe to contract and be "snuffed out"?

The balance between inward and outward pressure can be represented with

$$H = P - C \tag{6.11}$$

where P is the pressure of the energy-momentum contents of the virtual universe and C is the Casimir force due to the virtual vacuum energy embodied in the surface of the potential universe. There are three possibilities:

---

[29] Much of this chapter is based on Blaha (2022d) and (2023e).

[30] As we see later, the condition has the universe internal pressure rise with dimension. Thus at the equilibrium point the pressures balance. At higher dimensions the universe has higher internal pressure so it would be stable or expand.

[31] Subuniverses that might be created within the Parent universe have a different internal energy and a different Casimir force within the Parent universe due to a different internal Parent universe Casimir pressure on the subuniverse. Thus they may exist within the Parent.

[32] We showed previously that it leads to a set of expressions for important constants such as the origin of the power 2 $\cong e\pi/4$; the choice of 10 Cosmos spaces; the choice of a Parent universe of 18 space-time dimensions; and the choice of $c_g = 0.0785 \cong \pi^2/128$, which appears in a universal relation $kT = Mc_g$, is a new universal constant.

[33] In the *surface* of the virtual universe.

$$\begin{aligned} &H < 0 \quad \text{Casimir force would cause contraction to a point} \\ &H = 0 \quad \text{Balance of forces – stable} \\ &H > 0 \quad \text{Pressure causes (Hubble) expansion of universe} \end{aligned} \tag{6.12}$$

We wish to use H to determine the Cosmos Theory Physical spectrum. The spectrum has a lower Physical limit at r = 0 but no apparent upper limit. We now view H as determining the maximum dimension r_{Parent} of the Physical Cosmos spectrum.

We determine H for a universe of dimension r.

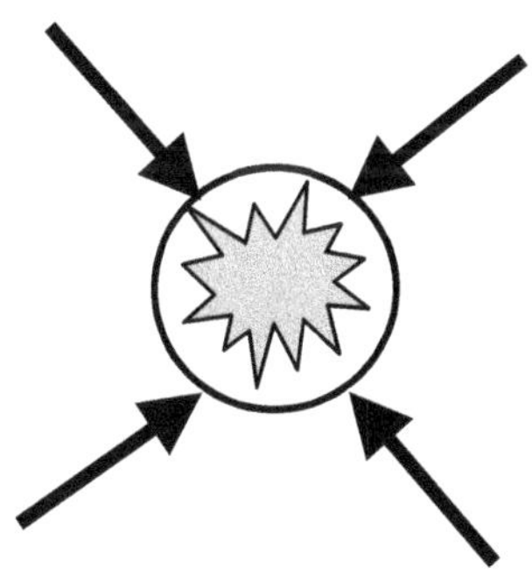

Figure 6.1. Schematic diagram of virtual universe's "vacuum" bubble showing the confining Casimir force of the exterior vacuum. This force is countered by internal fermion gas pressure symbolized by a jagged interior star.

6.6 Number of Fundamental Particle Species

Cosmos Theory specifies the number of fundamental fermion species N for a universe in a space of dimension r. The space is a combination of a HyperCosmos space and a 2nd Kind HyperCosmos space. Eq. 5.42a shows a combined state. So we now treat the number of fermion species as dependent on the angle θ:

$$N = \cos(\theta)\, 2^{r+4} + \sin(\theta)\, 2^{r+3} \tag{6.13}$$

We assume the number of fundamental fermions is the measure of the number of particles in a universe. The number of fermions can then be used to set P and determine H fixing the minimal universe of dimension r. In Cosmos Theory we have eq. 6.13 specifying the number of fermions for a universe of dimension r. In Sequence Theory we are led to quadruple the number of fermions into strata. Thus

$$N = \cos(\theta)\, 2^{r+6} + \sin(\theta)\, 2^{r+5} \tag{6.13}$$

The introduction of gambol fermions causes a further doubling of numbers to:

$$N_{Cosmos} = \cos(\theta)\, 2^{r+7} + \sin(\theta)\, 2^{r+6} \tag{6.14}$$

6.7 Conditions at the Consistency Point

We now consider an infinitesimal "box" containing the energy-momentum of a virtual universe. The box contains fermions with the initial features:

1. All fermions are massless.
2. All interactions are zero.
3. All baryon numbers are zero.

6.8 Thermodynamic Outward Pressure

We base the thermodynamic pressure on the Ideal Gas Law[34]

$$P = NkT/V \tag{6.15}$$

where P is the pressure, T is the temperature, k is Boltzmann's constant, and N is the number of particles in unit volume V.

Assuming the fermion gas occupies a spherical spatial ball with spatial volume

$$V_{r-1} = \pi^{(r-1)/2} a^{(r-1)} / \Gamma((r-1)/2 + 1) \tag{6.16}$$

with radius a in an r dimension space-time:

$$P = N\,\Gamma((r-1)/2 + 1)kT/(\pi^{(r-1)/2} a^{(r-1)}) \tag{6.17}$$

where $\Gamma(n)$ is the factorial function. The number of fermions N is given by eq. 6.14.

6.9 Casimir Effect Confining Force

This section calculates the force on a universe due to the Casimir effect.[35] We assume the universe is in a vacuum energy bubble in an r – 1 dimension spatial region of radius a. It generates a confining force on the surface of the universe. We start with the infinite Dirac fermion vacuum energy per unit volume:

$$U_\infty = \int_0^\infty d^{r-1}k\,(k^2 + m^2)^{1/2} \tag{6.18}$$

where m is the common fermion mass, which we take to be zero.

Due to the vacuum bubble, the vacuum energy changes to

$$U = U_\infty - U_{bubble} \tag{6.19}$$

[34] One might think of introducing more complex forms of gas laws based on the volume occupied by the gas particles and so on. However our assumption of zero interactions at the origin of the Parent universe eliminates the need for such "refinements." One might also thinking of introducing an attractive gravitation term in the consistency condition. Here again the assumption of no interactions at the origin point removes that concern.

[35] H. G. B. Casimir, Koninkl. Ned. Adak. Wetenschap. Proc. **51**, 793 (1948), S. K. Lamoreaux, Phys. Rev. Lett. **78**, 5 (1997), L. S. Brown and G. J. McClay, Phys. Rev. **184**, 1272 (1969), and references therein.

where the missing bubble vacuum energy per unit volume is

$$U_{bubble} = \int_d^\infty d^{r-1}k\,(k^2 + m^2)^{½} \qquad (6.20)$$

where d is the momentum space lower limit of the excluded vacuum energy within the bubble. The Casimir deficit is

$$U = \int_0^d d^{\,r-1}k\;k \qquad (6.21)$$

Assuming the momentum d corresponds to the spatial radius, a, of the bubble

$$a = 1/(mc) = 1/d \qquad (6.21a)$$

where c is the speed of light we find the energy deficit per unit volume is

$$U \cong \pi^{(r-1)/2}a^{-r}\,(r-1)/[r\Gamma((r-1)/2+1)] \qquad (6.22)$$

Multiplying by the volume of the bubble we find a total bubble energy:

$$U_{tot} \approx V_{r-1}\,U \qquad (6.23)$$

$$\approx \pi^{r-1}a^{-1}[(r-1)/r]/\Gamma((r-1)/2+1)^2 \qquad (6.24)$$

Then

$$\partial U_{tot}/\partial a \cong -\pi^{r-1}a^{-2}\,[(r-1)/r]/\Gamma((r-1)/2+1)^2 \qquad (6.25)$$

A bubble surface of radius a has the area

$$S_{r-2} = 2\pi^{(r-1)/2}\,a^{r-2}/\Gamma((r-1)/2) \qquad (6.26)$$

The Casimir force on the surface is therefore

$$C = (S_{r-2})^{-1}\,\partial U_{tot}/\partial a \qquad (6.27)$$
$$= -½\,\pi^{(r-1)/2}\,a^{-r}\,[(r-1)/r]\,\Gamma((r-1)/2)/\Gamma((r-1)/2+1)^2$$
$$= -\pi^{(r-1)/2}\,a^{-r}/[r\Gamma((r-1)/2+1)]$$

6.10 H Balance

We now set the outward pressure of the internal fermion "gas" equal to the inward Casimir vacuum force by eqs. 6.17 and 6.27:

$$H = P - C$$

Where

$$H = N_{Cosmos}\Gamma((r-1)/2+1)kT/(\pi^{(r-1)/2}a^{(r-1)}) - \pi^{(r-1)/2}a^{-r}/[r\Gamma((r-1)/2+1)] \tag{6.28}$$

Eliminating common factors in H we find the condition becomes

$$H' = N_{Cosmos}\, akT - \pi^{(r-1)}/[r\Gamma((r-1)/2+1)^2] \tag{6.29}$$

At the critical point of eq. 6.12 we see H' = 0 implies

$$N_{Cosmos} = \pi^{(r-1)}/[arkT\, \Gamma((r-1)/2+1)^2] \tag{6.30}$$

For collapse H' < 0

$$N_{Cosmos} < \pi^{(r-1)}/[arkT\, \Gamma((r-1)/2+1)^2] \tag{6.31}$$

For expansion H' > 0

$$N_{Cosmos} > \pi^{(r-1)}/[akTr\, (\Gamma((r-1)/2+1))^2] \tag{6.32}$$

6.11 N_{Cosmos} and H' as Functions of Dimension r and θ

The dependence of the critical point on θ suggests that we maximize the value of N_{Cosmos} as a function of θ to obtain the least value of r that meets the H' = 0 stability condition. The maximum N_{Cosmos}, after maximizing in θ, is

$$\begin{aligned} N_{CosmosMAX} &= 5 \cdot 2^{r+5} \cos(\theta) \\ &= 5 \cdot 2^{r+5} \end{aligned} \tag{6.33}$$

by setting cos(θ) = 1. We find that the choice of[36]

$$akT = 2.85 \times 10^{-12} \tag{6.34}$$

and eq. 6.33 give a critical point at dimension r = 18. Thus, at minimum, an r = 18 Parent universe can exist in equilibrium.

For even number dimensions below r = 18 we find H' is negative indicating a stable Parent universe could not exist. The Casimir force would prevent its existence. At r = 18 the pressure balances the Casimir force thus allowing an 18 dimension Parent universe within which child universes may be created. The r = 18 virtual universe is the minimal[37] *(in r) universe, and thus is the Parent universe dimension.*

The choice of cos θ = 1 implies the Parent universe has a HyperCosmos space dimension array.

On this basis we have Physical set of ten HyperCosmos spaces and ten 2nd Kind HyperCosmos spaces. We could create one or more 18 dimension Parent universes.

---

[36] The radius a of the Parent universe is viewed as very, very small as it appears near the "beginning" Mathematical "point" of the Cosmos Space. The temperature factor kT compensates by being extremely large. Together they result in the value of akT. Note akT is of the order of magnitude of $2^{39} = 1.81 \times 10^{-12}$.

[37] We require the dimension to be an even number for reasons stated previously.

Each Parent universe balances expansion and contraction. A Parent universe can contain subuniverses of lesser dimension since they have a less strong Casimir force with which to contend. *The subuniverses can have HyperCosmos space dimension arrays or 2nd Kind HyperCosmos space dimension arrays.*

6.12 Why a Parent Universe?

The existence of the Parent universe is simultaneous with the existence of the four components of Cosmos structure that we have considered previously. It has a dual role: it limits the number of Physically acceptable spaces to those with dimension less than 18; and it provides the arena for the dynamic creation of subuniverses.

The Parent universe arises when the virtual pressure of internal force balances the virtual Casimir pressure of the virtual vacuum. This balance is instantaneous with the instantaneity of the four components of Cosmos structure. It could not be otherwise since there is no mechanism for the creation of the Parent universe other than that of eqs. 6.1 and 6.2. They imply the joint instantaneity of the now *five* components of the Cosmos. Thus

The Parent universe is coexistent with the Cosmos.

In the following chapters we discuss aspects of the Parent universe and subuniverses.

6.12.1 Cosmos Structure Coordinates

The definition of time and spatial coordinates is necessarily based on the existence of tools that can measure their values from point to point. This fact is particularly evident when one imposes quantization to obtain Uncertainty Relations. Since no tools exist outside of universes one cannot define coordinates in any Physically meaningful way. Thus

There is no time outside of universes.
There are no coordinates outside of universes.

In the case of Cosmos structure there is an entity, the Cosmos Fractal Curve, which exists eternally in our view. We define a space based on its space-filling properties. We define a time in terms of points along the length of the Curve (as if it were straightened). Thus we obtain the coordinates of the two (plus one) dimension space generated from the Curve that is used to determine Cosmos structure. This equation appears in eq. 4.1. The time defined in this space has no connection with the time used within a universe.

Therefore the components of the Cosmos must be thought of as instantaneous. The instantaneity does not occur at any point in time.

An eternal instantaneity

6.12.1 Cosmos Structure Features

Cosmos structure has a number of important features:

1. It is defined in terms of spatial and time parameters in the two plus one dimension Cosmos Space generated from the Cosmos Fractal Curve.

2. It has no connection with universes except the Parent universe(s) where it participates in the definition of the Consistency Condition.

3. It is therefore eternal – coeval with the Parent universe(s). No beginning. No end. Unchanging!

4. It does not depend on any interaction beyond the scaling self-interactions appearing in eq. 4.1. Thus it is not subject to modification by other interactions in universes. The stability of Cosmos structure is based on the separation of Structure from Universe mass-energy as shown in our earlier books.

5. The Parent universe, with coordinates X_i in eq. 6.2, exists in the two plus one Cosmos Space. It has no interaction with other components although it has the dimension 18 specified by the Consistency Condition. If there were several Parent universes then they would also exist in isolation with no interactions between them. The Parent universe has a mass-energy and may possibly evolve. It is not created with a Big Bang since, like the other Cosmos components, it exists always as part of the Cosmos Simultaneity.[38]

6. The subuniverses of the Parent universe can be created with inter-universe interactions.[39] These interactions would also support dynamic interactions between subuniverses. The time scale of these interactions and of universe creation would presumably extremely large. The time would be measured by the internal Parent universe time. Within subuniverses there are internal times and interactions between energy and matter. In our universe the interactions would be specified by our Unified SuperStandard Theory (UST). Other universes would have analogous formulations of the UST.

7. Gravitation is universe-dependent. The two plus one dimension Cosmos Space has no gravitation since it has no mass-energy except for that of the Parent universe. Thus there is no gravitational interaction in Cosmos Space external to the Parent. The two plus one dimension Cosmos space is stable and fixed by our previous Cosmos structure considerations. Within the Parent universe there is a

---

[38] Described in our previous books.

[39] Possible models have been suggested in our previous books.

gravitation that could influence the motion of subuniverses. Each subuniverse has an internal gravitation interaction that influences the motions of its contents.

7. Hybrid Subuniverses

There has been much discussion of Multiverses in the literature and the Press. In this chapter we describe a type of Multiverse in which the dimension array partakes of the features of a HyperCosmos space and a 2nd Kind HyperCosmos space. We call a universe with this type of dimension array a Hybrid universe. The Parent universe described in previous chapters has a dimension array of the HyperCosmos type. The subuniverses described earlier may be of the Hybrid type or solely HyperCosmos type or 2nd HyperCosmos type.

7.1 Interpretation of the Hybrid Universe Quantum State

The state of a Hybrid universe of dimension r has a form similar to eq. 5.42a:

$$| r, \theta> = \cos(\theta) | r+4> + \sin(\theta) | r+3> \tag{7.1}$$

where θ is an angle between (but not equal to) zero and π/2. Each term has a corresponding dimension array. States of this Hybrid subuniverse will reflect the dimension arrays of each term. We view the separate terms as defining a subuniverse. We view | r, θ> as defining a composite (Multiverse?) Hybrid subuniverse. Note both subuniverses have the same space-time dimension.

These universes do not have explicit interactions with each other although interactions between them *might* be defined. We will assume they are independent and non-interacting, We also assume orthogonality:

$$< r+4| r+3> = 0 \tag{7.2}$$

7.2 Form of a Hybrid Dimension Array

The superposition of r+4 and r+3 dimension arrays in in eq. 7.1 requires some explanation. Each of the terms in | r, θ> can be viewed as specifying a dimension array for a universe. The hybrid dimension array can be physically viewed as combining a universe with a Dark sector and a universe with no Dark sector. The present quandary over the existence of a Dark sector for our universe makes the consideration of a hybrid universe worthwhile since our universe may be part of a hybrid universe.

The form of eq. 7.1 raises the issue of the interpretation of its dimension array which is necessarily composite. It specifies a type of "composition" of dimension arrays with sizes 2^{r+4} and 2^{r+3} if strata and gambol fermions are neglected. We label the items corresponding to the 2^{r+4} dimension array as "type 1." We label the items corresponding to the 2^{r+3} dimension array as "type 2."

The only reasonable interpretation appears to require the type 2 dimension array of size 2^{r+3} to expand with zero entries to the size 2^{r+4} for direct comparison. (The Normal fermions then become hybrid. These fermions have separate type 1 and type 2 interactions. The type 1 and type 2 symmetry groups are not the same. Since we see what appears to be fermions and symmetries unencumbered by a separation into types experimentally we expect the angle θ is either close to 0 or close to π/2 for <u>our</u> universe.)

In general, the Dark sector only has type 1 fermions and only type 1 interactions. We can depict the dimension arrays as in Fig. 7.1.

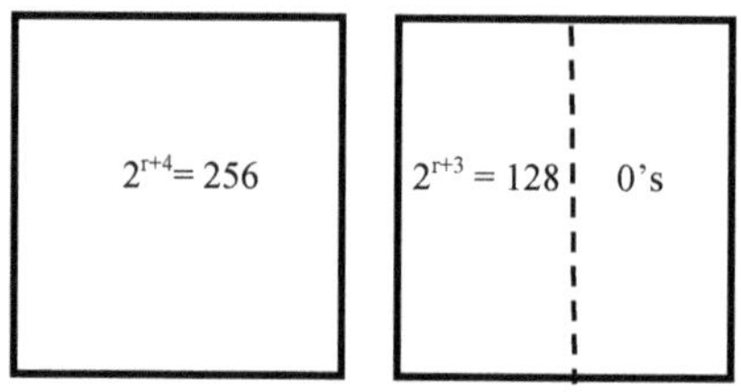

Figure 7.1. The type 1 and the expanded type 2 dimension arrays for <u>our</u> 4 dimension universe. The 256 dimension array has Normal and Dark sectors of 128 dimensions. We choose the right array to have a 128 dimension Normal sector and an expansion part of 128 dimensions filled with "zeros."

Each dimension array has a corresponding, but different, set of internal symmetries and interactions. We take *hybrid fermions* to coincide for the case of Normal fermions. The 2^{r+3} size dimension array for fermions does not have Dark fermions. Dark fermions are not hybrid.

Based on the above comments it appears *the angle* θ_r *in eq. 7.1 is close to 0 in our universe: We do not see hybrid fermion behavior.* Therefore we conclude (based on available evidence) in our universe:

1. The Normal fermions are predominantly 2^{r+3} dimension array fermions

2. Any Dark fermions of the 2^{r+4} size dimension array are suppressed by a factor of $\sin(\theta_r)$ in their UST Connection Group interactions with the Normal 2^{r+4} size dimension array fermions. Thus any interaction between Normal and Dark fermions is much suppressed by the factor $\sin(\theta_r) \approx 0$.

3. Dark fermions interact with each other via Dark interactions. Dark interactions do not support Dark fermion – Normal fermion interactions.

4. Therefore any Dark fermions (matter) are essentially independent of the Normal fermions as far as UST (Standard Model) interactions are concerned.

5. The only remaining interaction of note is Gravitation. We expect gravitation is the only significant interaction between Normal and Dark matter. The problem of the existence of Dark matter may be solved if our universe is a hybrid universe. Our mostly type 2 universe has Dark matter in its type 1 Dark sector. However there is no appreciable interaction between type 2

Normal matter and type 1 Dark matter except gravitation. And so experiment may "see" Dark matter effects gravitationally but not see Dark matter electromagnetically – the current experimental observations.

The above view of the implications of eq, 7.1 appear to be borne out by experiment. Therefore we are essentially in a universe with a 2nd Kind HyperCosmos space having a $2^{r+3} = 128$ size dimension array based on an angle θ that is, perhaps, ***slightly*** *greater than 0.*

8. The Eternal Cosmos

We have been developing a description of the Cosmos in Cosmos Theory for over six years. This description, being fairly complete, could be described as a simulation[41] of Reality to the extent that it mirrors Reality.

The task of this book is to unite the four components of Cosmos Theory using a Quantum unification approach. It has become clear that these components reflect the same reality when we consider the dynamics of the components in the self-generated two plus one dimension Cosmos Space. We have found the unified components are independent of the time in any universe. Timeless! They furnish the basis for the determination of a Parent universe(s) that contains the subuniverses of the Cosmos.

The Parent universe is itself timeless too. We see a Consistency Condition exists that fixes the dimension of the largest Physical Cosmos space, the Parent universe space (a HyperCosmos space) to be 18. The spectrum of Cosmos Theory spaces then has ten HyperCosmos spaces and ten 2nd Kind HyperCosmos spaces. There are other Physically interesting spaces that appear when the unification of symmetries is considered.

The role of the Consistency Condition, determining the number of Physical spaces, is based on the existence of a timeless Parent universe. We therefore increase the number of *timeless* components of Cosmos Theory to *five*. They are eternal. Although we describe Cosmos Theory in a "time ordered" sequence the components, and their interrelations, are instantaneously "implemented." When does this happen? In the absence of a time it happens always - eternally.

The structure of the five component Cosmos Theory is not subject to interactions of any kind. They are dynamically determined by the Lagrangian and extended Master Equation presented in preceding chapters. Thus their constitution is stable and unchanging.

We separated the structure of Cosmos Theory from mass-energy entities. This separation enabled us to specify a stable, unchanging structure while allowing the dynamic evolution of mass-energy entities – subuniverses. We quantized subuniverses. Their quantum fields were discussed in an earlier chapter.

Within subuniverses particles and symmetries exist that determine their evolution. In the case of our universe we developed a Unified SuperStandard Theory over a period of over ten years that describes Elementary Particles. It is a major outgrowth of The Standard Model.

We now have a Cosmos Theory based on a Quantum unification of its five components. The theory emerges from a mathematical point – evolves to a one dimension line – that is used to construct a space-filled two dimension square – that becomes Cosmos space when we adjoin a "time" dimension that is simply the running length of the Cosmos Fractal Curve. In this Cosmos Space we define the dynamics of the HyperCosmos spaces spectrum, the Master Equation, and its eigenvalues (that set the size of space dimension arrays.) Cosmos Theory begins with the Cosmos Fractal Curve!

Based on this development we have dimension arrays specifying the fundamental fermion spectrums, the symmetries, and the consequent interactions of particles and energies. The interactions, fermion masses and coupling constants follow from sequences specified by the Master Equation.

The thrust of this book is to weld the components and the constructs that they generate (subuniverses and their contents) into a stream that does not necessarily require external intervention through an agency such as a Prime Mover.

The question of "Why" is partially answered. Some remaining issues are: the origin ("spark") for the "creation" of the Cosmos Fractal Curve – the source of it all, and the origin of the hierarchy of subuniverses of the Parent universe. This second question could be resolved through a theory of Parent and subuniverse interactions.

Alea Iacta Est

REFERENCES

Akhiezer, N. I., Frink, A. H. (tr), 1962, *The Calculus of Variations* (Blaisdell Publishing, New York, 1962).
Bjorken, J. D., Drell, S. D., 1964, *Relativistic Quantum Mechanics* (McGraw-Hill, New York, 1965).
Bjorken, J. D., Drell, S. D., 1965, *Relativistic Quantum Fields* (McGraw-Hill, New York, 1965).
Blaha, S., 1995, *C++ for Professional Programming* (International Thomson Publishing, Boston, 1995).
_______, 1998, *Cosmos and Consciousness* (Pingree-Hill Publishing, Auburn, NH, 1998 and 2002).
_______, 2002, *A Finite Unified Quantum Field Theory of the Elementary Particle Standard Model and Quantum Gravity Based on New Quantum Dimensions™ & a New Paradigm in the Calculus of Variations* (Pingree-Hill Publishing, Auburn, NH, 2002).
_______, 2004, *Quantum Big Bang Cosmology: Complex Space-time General Relativity, Quantum Coordinates™ Dodecahedral Universe, Inflation, and New Spin 0, ½, 1 & 2 Tachyons & Imagyons* (Pingree-Hill Publishing, Auburn, NH, 2004).
______, *2005a, Quantum Theory of the Third Kind: A New Type of Divergence-free Quantum Field Theory Supporting a Unified Standard Model of Elementary Particles and Quantum Gravity based on a New Method in the Calculus of Variations* (Pingree-Hill Publishing, Auburn, NH, 2005).
______, 2005b, *The Metatheory of Physics Theories, and the Theory of Everything as a Quantum Computer Language* (Pingree-Hill Publishing, Auburn, NH, 2005).
______, 2005c, *The Equivalence of Elementary Particle Theories and Computer Languages: Quantum Computers, Turing Machines, Standard Model, Superstring Theory, and a Proof that Gödel's Theorem Implies Nature Must Be Quantum* (Pingree-Hill Publishing, Auburn, NH, 2005).
______, 2006a, *The Foundation of the Forces of Nature* (Pingree-Hill Publishing, Auburn, NH, 2006).
______, 2006b, *A Derivation of ElectroWeak Theory based on an Extension of Special Relativity; Black Hole Tachyons; & Tachyons of Any Spin.* (Pingree-Hill Publishing, Auburn, NH, 2006).
_______, 2007a, *Physics Beyond the Light Barrier: The Source of Parity Violation, Tachyons, and A Derivation of Standard Model Features* (Pingree-Hill Publishing, Auburn, NH, 2007).
______, 2007b, *The Origin of the Standard Model: The Genesis of Four Quark and Lepton Species, Parity Violation, the ElectroWeak Sector, Color SU(3), Three Visible Generations of Fermions, and One Generation of Dark Matter with Dark Energy* (Pingree-Hill Publishing, Auburn, NH, 2007).
______, *2008a, A Direct Derivation of the Form of the Standard Model From GL(16) (Pingree-Hill Publishing, Auburn, NH, 2008).*
______, 2008b, *A Complete Derivation of the Form of the Standard Model With a New Method to Generate Particle Masses Second Edition* (Pingree-Hill Publishing, Auburn, NH, 2008)
______, 2009, *The Algebra of Thought & Reality: The Mathematical Basis for Plato's Theory of Ideas, and Reality Extended to Include A Priori Observers and Space-Time Second Edition* (Pingree-Hill Publishing, Auburn, NH, 2009).
______, 2009a, *Bright Stars, Bright Universe* (Pingree-Hill Publishing, Auburn, NH, 2008)
______, 2010a, *Operator Metaphysics: A New Metaphysics Based on a New Operator Logic and a New Quantum Operator Logic that Lead to a Mathematical Basis for Plato's Theory of Ideas and Reality* (Pingree-Hill Publishing, Auburn, NH, 2010).
______, 2010b, *The Standard Model's Form Derived from Operator Logic, Superluminal Transformations and GL(16)* (Pingree-Hill Publishing, Auburn, NH, 2010).
______, 2010c, *SuperCivilizations: Civilizations as Superorganisms* (McMann-Fisher Publishing, Auburn, NH, 2010).
______, 2011a, *21st Century Natural Philosophy Of Ultimate Physical Reality* (McMann-Fisher Publishing, Auburn, NH, 2011).
______, 2011b, *All the Universe! Faster Than Light Tachyon Quark Starships & Particle Accelerators with the LHC as a Prototype Starship Drive Scientific Edition* (Pingree-Hill Publishing, Auburn, NH, 2011).
______, 2011c, *From Asynchronous Logic to The Standard Model to Superflight to the Stars* (Blaha Research, Auburn, NH, 2011).

______, 2012a, *From Asynchronous Logic to The Standard Model to Superflight to the Stars volume 2: Superluminal CP and CPT, U(4) Complex General Relativity and The Standard Model, Complex Vierbein General Relativity, Kinetic Theory, Thermodynamics* (Blaha Research, Auburn, NH, 2012).

______, 2012b, *Standard Model Symmetries, And Four And Sixteen Dimension Complex Relativity; The Origin Of Higgs Mass Terms* (Blaha Reasearch, Auburn, NH, 2012).

______, 2013a, *Multi-Stage Space Guns, Micro-Pulse Nuclear Rockets, and Faster-Than-Light Quark-Gluon Ion Drive Starships* (Blaha Research, Auburn, NH, 2013).

______, 2013b, *The Bridge to Dark Matter; A New Sibling Universe; Dark Energy; Inflatons; Quantum Big Bang; Superluminal Physics; An Extended Standard Model Based on Geometry* (Blaha Reasearch, Auburn, NH, 2013).

______, 2014a, *Universes and Megaverses: From a New Standard Model to a Physical Megaverse; The Big Bang; Our Sibling Universe's Wormhole; Origin of the Cosmological Constant, Spatial Asymmetry of the Universe, and its Web of Galaxies; A Baryonic Field between Universes and Particles; Megaverse Extended Wheeler-DeWitt Equation* (Blaha Reasearch, Auburn, NH, 2014).

______, 2014b, *All the Megaverse! Starships Exploring the Endless Universes of the Cosmos Using the Baryonic Force* (Blaha Research, Auburn, NH, 2014).

______, 2014c, *All the Megaverse! II Between Megaverse Universes: Quantum Entanglement Explained by the Megaverse Coherent Baryonic Radiation Devices – PHASERs Neutron Star Megaverse Slingshot Dynamics Spiritual and UFO Events, and the Megaverse Microscopic Entry into the Megaverse* (Blaha Research, Auburn, NH, 2014).

______, 2015a, *PHYSICS IS LOGIC PAINTED ON THE VOID: Origin of Bare Masses and The Standard Model in Logic, U(4) Origin of the Generations, Normal and Dark Baryonic Forces, Dark Matter, Dark Energy, The Big Bang, Complex General Relativity, A Megaverse of Universe Particles* (Blaha Research, Auburn, NH, 2015).

______, 2015b, *PHYSICS IS LOGIC Part II: The Theory of Everything, The Megaverse Theory of Everything, U(4)⊗U(4) Grand Unified Theory (GUT), Inertial Mass = Gravitational Mass, Unified Extended Standard Model and a New Complex General Relativity with Higgs Particles, Generation Group Higgs Particles* (Blaha Research, Auburn, NH, 2015).

____, 2015c, *The Origin of Higgs ("God") Particles and the Higgs Mechanism: Physics is Logic III, Beyond Higgs – A Revamped Theory With a Local Arrow of Time, The Theory of Everything Enhanced, Why Inertial Frames are Special, Universes of the Mind* (Blaha Research, Auburn, NH, 2015).

______, 2015d, *The Origin of the Eight Coupling Constants of The Theory of Everything: U(8) Grand Unified Theory of Everything (GUTE), S^8 Coupling Constant Symmetry, Space-Time Dependent Coupling Constants, Big Bang Vacuum Coupling Constants, Physics is Logic IV* (Blaha Research, Auburn, NH, 2015).

______, 2016a, *New Types of Dark Matter, Big Bang Equipartition, and A New U(4) Symmetry in the Theory of Everything: Equipartition Principle for Fermions, Matter is 83.33% Dark, Penetrating the Veil of the Big Bang, Explicit QFT Quark Confinement and Charmonium, Physics is Logic V* (Blaha Research, Auburn, NH, 2016).

______, 2016b, *The Periodic Table of the 192 Quarks and Leptons in The Theory of Everything: The U(4) Layer Group, Physics is Logic VI* (Blaha Research, Auburn, NH, 2016).

______, 2016c, *New Boson Quantum Field Theory, Dark Matter Dynamics, Dark Matter Fermion Layer Mixing, Genesis of Higgs Particles, New Layer Higgs Masses, Higgs Coupling Constants, Non-Abelian Higgs Gauge Fields, Physics is Logic VII* (Blaha Research, Auburn, NH, 2016).

______, 2016d, *Unification of the Strong Interactions and Gravitation: Quark Confinement Linked to Modified Short-Distance Gravity; Physics is Logic VIII* (Blaha Research, Auburn, NH, 2016).

______, 2016e, *MoND: Unification of the Strong Interactions and Gravitation II, Quark Confinement Linked to Large-Scale Gravity, Physics is Logic IX* (Blaha Research, Auburn, NH, 2016).

______, 2016f, *CQ Mechanics: A Unification of Quantum & Classical Mechanics, Quantum/Semi-Classical Entanglement, Quantum/Classical Path Integrals, Quantum/Classical Chaos* (Blaha Research, Auburn, NH, 2016).

______, 2016g, *GEMS Unified Gravity, ElectroMagnetic and Strong Interactions: Manifest Quark Confinement, A Solution for the Proton Spin Puzzle, Modified Gravity on the Galactic Scale* (Pingree Hill Publishing, Auburn, NH, 2016).

______, 2016h, *Unification of the Seven Boson Interactions based on the Riemann-Christoffel Curvature Tensor* (Pingree Hill Publishing, Auburn, NH, 2016).

______, 2017a, *Unification of the Eleven Boson Interactions based on 'Rotations of Interactions'* (Pingree Hill Publishing, Auburn, NH, 2017).

______, 2017b, *The Origin of Fermions and Bosons, and Their Unification* (Pingree Hill Publishing, Auburn, NH, 2017).

______, 2017c, *Megaverse: The Universe of Universes* (Pingree Hill Publishing, Auburn, NH, 2017).

______, 2017d, *SuperSymmetry and the Unified SuperStandard Model* (Pingree Hill Publishing, Auburn, NH, 2017).

______, 2017e, *From Qubits to the Unified SuperStandard Model with Embedded SuperStrings: A Derivation* (Pingree Hill Publishing, Auburn, NH, 2017).
______, 2017f, *The Unified SuperStandard Model in Our Universe and the Megaverse: Quarks, ...* , (Pingree Hill Publishing, Auburn, NH, 2017).
______, 2018a, *The Unified SuperStandard Model and the Megaverse SECOND EDITION A Deeper Theory based on a New Particle Functional Space that Explicates Quantum Entanglement Spookiness (Volume 1)* (Pingree Hill Publishing, Auburn, NH, 2018).
______, 2018b, *Cosmos Creation: The Unified SuperStandard Model, Volume 2, SECOND EDITION* (Pingree Hill Publishing, Auburn, NH, 2018).
______, 2018c, *God Theory (*Pingree Hill Publishing, Auburn, NH, 2018).
______, 2018d, *Immortal Eye: God Theory: Second Edition* (Pingree Hill Publishing, Auburn, NH, 2018).
______, 2018e, *Unification of God Theory and Unified SuperStandard Model THIRD EDITION* (Pingree Hill Publishing, Auburn, NH, 2018).
______, 2019a, *Calculation of: QED α = 1/137, and Other Coupling Constants of the Unified SuperStandard Theory* (Pingree Hill Publishing, Auburn, NH, 2019).
______, 2019b, *Coupling Constants of the Unified SuperStandard Theory SECOND EDITION* (Pingree Hill Publishing, Auburn, NH, 2019).
______, 2019c, *New Hybrid Quantum Big_Bang–Megaverse_Driven Universe with a Finite Big Bang and an Increasing Hubble Constant* (Pingree Hill Publishing, Auburn, NH, 2019).
______, 2019d, *The Universe, The Electron and The Vacuum* (Pingree Hill Publishing, Auburn, NH, 2019).
______, 2019e, *Quantum Big Bang – Quantum Vacuum Universes (Particles)* (Pingree Hill Publishing, Auburn, NH, 2019).
______, 2019f, *The Exact QED Calculation of the Fine Structure Constant Implies ALL 4D Universes have the Same Physics/Life Prospects* (Pingree Hill Publishing, Auburn, NH, 2019).
______, 2019g, *Unified SuperStandard Theory and the SuperUniverse Model: The Foundation of Science* (Pingree Hill Publishing, Auburn, NH, 2019).
______, 2020a, *Quaternion Unified SuperStandard Theory (The QUeST) and Megaverse Octonion SuperStandard Theory (MOST)* (Pingree Hill Publishing, Auburn, NH, 2020).
______, 2020b, *United Universes Quaternion Universe - Octonion Megaverse* (Pingree Hill Publishing, Auburn, NH, 2020).
______, 2020c, *Unified SuperStandard Theories for Quaternion Universes & The Octonion Megaverse* (Pingree Hill Publishing, Auburn, NH, 2020).
______, 2020d, *The Essence of Eternity: Quaternion & Octonion SuperStandard Theories* (Pingree Hill Publishing, Auburn, NH, 2020).
______, 2020e, *The Essence of Eternity II* (Pingree Hill Publishing, Auburn, NH, 2020).
______, 2020f, *A Very Conscious Universe* (Pingree Hill Publishing, Auburn, NH, 2020).
______, 2020g, *Hypercomplex Universe* (Pingree Hill Publishing, Auburn, NH, 2020).
______, 2020h, *Beneath the Quaternion Universe* (Pingree Hill Publishing, Auburn, NH, 2020).
______, 2020i, *Why is the Universe Real? From Quaternion & Octonion to Real Coordinates* (Pingree Hill Publishing, Auburn, NH, 2020).
______, 2020j, *The Origin of Universes: of Quaternion Unified SuperStandard Theory (QUeST); and of the Octonion Megaverse (UTMOST)* (Pingree Hill Publishing, Auburn, NH, 2020).
______, 2020k, *The Seven Spaces of Creation: Octonion Cosmology* (Pingree Hill Publishing, Auburn, NH, 2020).
______, 2020l, *From Octonion Cosmology to the Unified SuperStandard Theory of Particles* (Pingree Hill Publishing, Auburn, NH, 2020).
______, 2021a, *Pioneering the Cosmos* (Pingree Hill Publishing, Auburn, NH, 2021).
______, 2021b, *Pioneering the Cosmos II* (Pingree Hill Publishing, Auburn, NH, 2021).
______, 2021c, *Beyond Octonion Cosmology* (Pingree Hill Publishing, Auburn, NH, 2021).
______, 2021d, *Universes are Particles* (Pingree Hill Publishing, Auburn, NH, 2021).
______, 2021e, *Octonion-like dna-based life, Universe expansion is decay, Emerging New Physics* (Pingree Hill Publishing, Auburn, NH, 2021).
______, 2021f, *The Science of Creation New Quantum Field Theory of Spaces* (Pingree Hill Publishing, Auburn, NH, 2021).
______, 2021g, *Quantum Space Theory With Application to Octonion Cosmology & Possibly To Fermionic Condensed Matter* (Pingree Hill Publishing, Auburn, NH, 2021).
______, 2021h, *21st Century Natural Philosophy of Octonion Cosmology , and Predestination, Fate, and Free Will* (Pingree Hill Publishing, Auburn, NH, 2021).

______, 2021i, *Beyond Octonion Cosmology II : Origin of the Quantum; A New Generalized Field Theory (GiFT); A Proof of the Spectrum of Universes; Atoms in Higher Universes* (Pingree Hill Publishing, Auburn, NH, 2021).
______, 2021j, *Integration of General Relativity and Quantum Theory: Octonion Cosmology, GiFT, Creation/Annihilation Spaces CASe, Reduction of Spaces to a Few Fermions and Symmetries in Fundamental Frames* (Pingree Hill Publishing, Auburn, NH, 2021).
______, 2022a, *New View of Octonion Cosmology Based on the Unification of General Relativity and Quantum Theory* (Pingree Hill Publishing, Auburn, NH, 2022).
______, 2022b, *The Dust Beneath Hypercomplex Cosmology* (Pingree Hill Publishing, Auburn, NH, 2022).
______, 2022c, *Passing Through Nature to Eternity: ProtoCosmos, HyperCosmos, Unified SuperStandard Theory* (Pingree Hill Publishing, Auburn, NH, 2022).
______, 2022d, *HyperCosmos Fractionation and Fundamental Reference Frame Based Unification: Particle Inner Space Basis of Parton and Dual Resonance Models* (Pingree Hill Publishing, Auburn, NH, 2022).
______, 2022e, *A New UniDimension ProtoCosmos and SuperString F-Theory Relation to the HyperCosmos* (Pingree Hill Publishing, Auburn, NH, 2022).
______, 2022f, *The Cosmic Panorama: ProtoCosmos, HyperCosmos, Unified SuperStandard Theory (UST) Derivation* (Pingree Hill Publishing, Auburn, NH, 2022).
______, 2022g, *Ultimate Origin: ProtoCosmos and HyperCosmos* (Pingree Hill Publishing, Auburn, NH, 2022).
______, 2023a, *UltraUnification and the Generation of the Cosmos* (Pingree Hill Publishing, Auburn, NH, 2023).
______, 2023b, *God and and Cosmos Theory* (Pingree Hill Publishing, Auburn, NH, 2023).
______, 2023c, *A New Completely Geometric SU(8) Cosmos Theory; New PseudoFermion Fields; Fibonacci-like Dimension Arrays; Ramsey Number Approximation* (Pingree Hill Publishing, Auburn, NH, 2023).
______, 2023d, *Newton's Apple is Now the Fermion* (Pingree Hill Publishing, Auburn, NH, 2023).
______, 2023e, *Cosmos Theory: The Sub-Particle Gambol Model* (Pingree Hill Publishing, Auburn, NH, 2023).
______, 2024a, *Cosmos-Universe-Particle-Gambol Theory* (Pingree Hill Publishing, Auburn, NH, 2024).
______, 2024b, *Fractal Cosmos Theory* (Pingree Hill Publishing, Auburn, NH, 2024).
______, 2024c, *Fractal Cosmic Curve: Tensor-Based CosmosTheory* (Pingree Hill Publishing, Auburn, NH, 2024).
______, 2024d, *The Eternal Form of Cosmos Theory* (Pingree Hill Publishing, Auburn, NH, 2024).
______, 2024e, *The Eternal Form of Cosmos Theory Third Edition* (Pingree Hill Publishing, Auburn, NH, 2024).
______, 2024f, *Fundamental Constants of Cosmos Theory and The Standard Model* (Pingree Hill Publishing, Auburn, NH, 2024).
______, 2024g, *Quark, Lepton, W and Z Masses of Cosmos Theory and The Standard Model* (Pingree Hill Publishing, Auburn, NH, 2024).
______, 2024h, *Geometric Cosmos Geometric Universe* (Pingree Hill Publishing, Auburn, NH, 2024).
______, 2024i, *Particles and Universes of Cosmos Theory* (Pingree Hill Publishing, Auburn, NH, 2024).
______, 2024j, *Unification of the Subluminal and the Superluminal in Cosmos Theory* (Pingree Hill Publishing, Auburn, NH, 2024).
______, 2024k, *The Dawn of Dynamic Cosmos Dimension Arrays* (Pingree Hill Publishing, Auburn, NH, 2024).
______, 2024ℓ, *Structure and Dynamics of Cosmos Theory and the Unified SuperStandard Theory* (Pingree Hill Publishing, Auburn, NH, 2024).
______, 2025a, *Black Holes, White Holes, and Superluminal Starship* (Pingree Hill Publishing, Auburn, NH, 2025).
______, 2025b, *The Sequence Theory Foundation of Cosmos Theory and Gambol Theory* (Pingree Hill Publishing, Auburn, NH, 2025).
______, 2025c, *BEYOND TIME AND SPACE Fermion Strata & Consistency Conditions for Sequences, Cosmos Spaces and Gambols* (Pingree Hill Publishing, Auburn, NH, 2025).
______, 2025d, *The C o s m o s as a Physics Organism* (Blaha Research, Auburn, NH, 2025).
______, 2025e, *The Root of Reality* (Blaha Research, Auburn, NH, 2025).
______, 2025f, *The Fractal Cosmos* (Blaha Research, Auburn, NH, 2025).
______, 2025g, *The Fractal Cosmos II* (Blaha Research, Auburn, NH, 2025).
______, 2025h, *The Fractal Cosmos III* (Blaha Research, Auburn, NH, 2025).
Eddington, A. S., 1952, *The Mathematical Theory of Relativity* (Cambridge University Press, Cambridge, U.K., 1952).
Fant, Karl M., 2005, *Logically Determined Design: Clockless System Design With NULL Convention Logic* (John Wiley and Sons, Hoboken, NJ, 2005).
Feinberg, G. and Shapiro, R., 1980, *Life Beyond Earth: The Intelligent Earthlings Guide to Life in the Universe* (William Morrow and Company, New York, 1980).
Gelfand, I. M., Fomin, S. V., Silverman, R. A. (tr), 2000, *Calculus of Variations* (Dover Publications, Mineola, NY, 2000).

Giaquinta, M., Modica, G., Souchek, J., 1998, *Cartesian Coordinates in the Calculus of Variations* Volumes I and II (Springer-Verlag, New York, 1998).
Giaquinta, M., Hildebrandt, S., 1996, *Calculus of Variations* Volumes I and II (Springer-Verlag, New York, 1996).
Gradshteyn, I. S. and Ryzhik, I. M., 1965, *Table of Integrals, Series, and Products* (Academic Press, New York, 1965).
Heitler, W., 1954, *The Quantum Theory of Radiation* (Claendon Press, Oxford, UK, 1954).
Huang, Kerson, 1992, *Quarks, Leptons & Gauge Fields 2nd Edition* (World Scientific Publishing Company, Singaporc, 1992).
Jost, J., Li-Jost, X., 1998, *Calculus of Variations* (Cambridge University Press, New York, 1998).
Kaku, Michio, 1993, *Quantum Field Theory*, (Oxford University Press, New York, 1993).
Kirk, G. S. and Raven, J. E., 1962, *The Presocratic Philosophers* (Cambridge University Press, New York, 1962).
Landau, L. D. and Lifshitz, E. M., 1987, *Fluid Mechanics 2nd Edition*, (Pergamon Press, Elmsford, NY, 1987).
Rescher, N., 1967, *The Philosophy of Leibniz* (Prentice-Hall, Englewood Cliffs, NJ, 1967).
Riesz, Frigyes and Sz.-Nagy, Béla, 1990, *Functional Analysis* (Dover Publications, New York, 1990).
Sakurai, J. J., 1964, *Invariance Principles and Elementary Particles* (Princeton University Press, Princeton, NJ, 1964).
Weinberg, S., 1972, *Gravitation and Cosmology* (John Wiley and Sons, New York, 1972).
Weinberg, S., 1995, *The Quantum Theory of Fields Volume I* (Cambridge University Press, New York, 1995).

INDEX

2nd Kind HyperCosmos, 7, 1, 4, 6, 9, 10, 11, 16, 35, 36, 40, 41, 43, 46, 51, 53, 55
Black Hole, 57
Bruno, 1
Casimir, 5, 7, 42, 43, 44, 45, 46, 47, 68
Classical, 58
coalescence, 17
confinement, 68
Consistency Condition, 5, 1, 39, 41, 42, 48, 50, 55
Cosmological Constant, 58
Cosmos, 59, 66
Cosmos Fractal Curve, 27
Cosmos Theory, 60
Creation, 59
Dark Energy, 57
Dark Matter, 57, 58
deep inelastic, 66
divergences, 65
electron, 66
ElectroWeak, 57
fine structure constant, 66
Fractal, 60
Fractal Cosmic Curve, 60
Fractal Cosmos Curve, 10
Gambol Theory, 4, 60
Gaussian, 31, 32, 36, 37
Generation, 57, 58, 60
Generation Group, 58
Geometry, 58
Gravitation, 52
hadron, 68
hadron scattering, 68
harmonic oscillator, 65
hybrid dimension array, 51
hybrid fermions, 52
Hybrid universe, 51
HyperCosmos, 4, 60, 68
hypersphere, 8
HyperUnification, 68
HyperUnification Spaces, 68
instantaneity, 33, 47
interactions, 65
ISIS, 67
Kepler, 1
kT, 46
Lebesgue, 9
Leibniz, 61
LHC, 57
Limos, 68
Many Worlds Theory, 1
Megaverse, 4, 58, 59, 68
Multiverse, 51
Music of the Hyperspheres, 10
paradoxes, 65
Parent universe, 1, 2, 3, 4, 18, 33, 39, 41, 42, 44, 46, 47, 48, 50, 51, 55, 56
Parity Violation, 57
perturbation theory, 66
Physical Reality, 57
Planckian, 68
Plato's Theory of Ideas, 57
Prime Mover, 1, 2, 3, 17, 56
ProtoCosmos, 4, 60, 68
PseudoFermion, 60
QED, 4, 59, 66
Quantum, 59, 65, 67
Quantum Dimensions, 57
Quantum Entanglement, 58
Quantum Gravity, 57
quark, 65
Riemann zeta function, 32
scaling, 66
scaling potential, 12, 19, 22, 24
Sequence Theory, 4, 60
simulation, 1, 50, 55
space-time, 66, 68
Special Relativity, 57
spin, 65
Standard Model, 57, 58, 60, 65, 66, 67

subuniverses, 1, 2, 3, 4, 39, 41, 42, 47, 48, 49, 51, 55, 56
Thermodynamics, 58
time, 57, 65, 66, 68
timeless, 1, 33, 55
tweak, 17
Two Tier Theory, 11
UltraUnification, 60
UniDimension ProtoCosmos, 60
Unification, 4, 58, 59, 60
Unified SuperStandard Model, 4, 59
Unified SuperStandard Theory, 4, 59, 60, 66, 67
unit square, 7
universe, 65, 66, 67, 68
universes, 66, 67, 68
Universes, 4, 58, 59, 60
UST, 4, 48, 52, 60, 67
Web of Galaxies, 58

About the Author

Stephen Blaha is a well-known Physicist and Man of Letters with interests in Science, Society and civilization, the Arts, and Technology. He had an Alfred P. Sloan Foundation scholarship in college. He received his Ph.D. in Physics from Rockefeller University. He has served on the faculties of several major universities. He was also a Member of the Technical Staff at Bell Laboratories, a manager at the Boston Globe Newspaper, a Director at Wang Laboratories, and President of Blaha Software Inc. and of Janus Associates Inc. (NH).

Among other achievements he was a co-discoverer of the "r potential" for heavy quark binding developing the first (and still the only demonstrable) non-Aeolian gauge theory with an "r" potential; first suggested the existence of topological structures in superfluid He-3; first proposed Yang-Mills theories would appear in condensed matter phenomena with non-scalar order parameters; first developed a grammar-based formalism for quantum computers and applied it to elementary particle theories; first developed a new form of quantum field theory without divergences (thus solving a major 60 year old problem that enabled a unified theory of the Standard Model and Quantum Gravity without divergences to be developed); first developed a formulation of complex General Relativity based on analytic continuation from real space-time; first developed a generalized non-homogeneous Robertson-Walker metric that enabled a quantum theory of the Big Bang to be developed without singularities at t = 0; first generalized Cauchy's theorem and Gauss' theorem to complex, curved multi-dimensional spaces; received Honorable Mention in the Gravity Research Foundation Essay Competition in 1978; first developed a physically acceptable theory of faster-than-light particles; first derived a composition of extremums method in the Calculus of Variations; first quantitatively suggested that inflationary periods in the history of the universe were not needed; first proved Gödel's Theorem implies Nature must be quantum; provided a new alternative to the Higgs Mechanism, and Higgs particles, to generate masses; first showed how to resolve logical paradoxes including Gödel's Undecidability Theorem by developing Operator Logic and Quantum Operator Logic; first developed a quantitative harmonic oscillator-like model of the life cycle, and interactions, of civilizations; first showed how equations describing superorganisms also apply to civilizations. A recent book shows his theory applies successfully to the past 14 years of history and to *new* archaeological data on Andean and Mayan civilizations as well as Early Anatolian and Egyptian civilizations.

He first developed an axiomatic derivation of the form of The Standard Model from geometry – space-time properties – The Unified SuperStandard Model. It unifies all the known forces of Nature. It also has a Dark Matter sector that includes a Dark ElectroWeak sector with Dark doublets and Dark gauge interactions. It uses quantum coordinates to remove infinities that crop up in most interacting quantum field theories and additionally to remove the infinities that appear in the Big Bang and generate inflationary growth of the universe. It shows gravity has a MOND-like form without sacrificing Newton's Laws. It relates the interactions of the MOND-like sector of gravity with the r-potential of Quark Confinement. The axioms of the theory lead to the question of their origin. We suggest in the preceding edition of this book it can be attributed to an entity with God-like properties. We explore these properties in "God Theory" and show they predict that the Cosmos exists forever although individual universes (or incarnations of our universe) "come and go." Several other important results emerge from God Theory such a functionally triune God. The Unified SuperStandard Theory has many other important parts described in the Current Edition of *The Unified SuperStandard Theory* and expanded in subsequent volumes.

Blaha has had a major impact on a succession of elementary particle theories: his Ph.D. thesis (1970), and papers, showed that quantum field theory calculations to all orders in ladder approximations could not give scaling deep inelastic electron-nucleon scattering. He later showed the eigenvalue equation for the fine structure constant α in Johnson-Baker-Willey QED had a zero at α = 1 not 1/137 by solving the Schwinger-Dyson equations to all orders in an approximation that agreed with exact results to 4th order in α thus ending interest in this theory. In 1979 at Professor Ken Johnson's (MIT) suggestion he calculated the proton-neutron mass difference in the MIT bag model and found the result had the wrong sign reducing interest in the bag model. These results all appear in Physical Review papers. In the 2000's he repeatedly pointed out the shortcomings of SuperString theory and showed that The Standard Model's form could be derived from space-time geometry by an extension of Lorentz transformations to faster than light transformations. This deeper space-time basis greatly increases the possibility that it is part of THE fundamental theory. Recently, Blaha showed that the Weak interactions differed significantly from the Strong, electromagnetic and gravitation interactions in important respects while these interactions had similar features, and suggested that ElectroWeak theory, which is essentially a glued union of the Weak interactions and Electromagnetism, possibly modulo unknown Higgs particle features, be replaced by a unified theory of the other interactions combined with a stand-alone Weak interaction theory. Blaha also showed that, if Charmonium calculations are taken seriously, the Strong interaction coupling constant is only a factor of five larger than the electromagnetic coupling constant, and thus Strong interaction perturbation theory would make sense and yield physically meaningful results.

In graduate school (1965-71) he wrote substantial papers in elementary particles and group theory: The Inelastic E- P Structure Functions in a Gluon Model. Phys. Lett. B40:501-502,1972; Deep-Inelastic E-P Structure Functions In A Ladder Model With Spin 1/2 Nucleons, Phys.Rev. D3:510-523,1971; Continuum Contributions To The Pion Radius, Phys. Rev. 178:2167-2169,1969; Character Analysis of U(N) and SU(N), J. Math. Phys. 10, 2156 (1969); and The Calculation of the Irreducible Characters of the Symmetric Group in Terms of the Compound Characters, (Published as Blaha's Lemma in D. E. Knuth's book: *The Art of Computer Programming Vols. 1 – 4*).

In the early 1980's Blaha was also a pioneer in the development of UNIX for financial, scientific and Internet applications: benchmarked UNIX versions showing that block size was critical for UNIX performance, developing financial modeling software, starting database benchmarking comparison studies, developing Internet-like UNIX networking (1982) and developing a hybrid shell programming technique (1982) that was a precursor to the PERL programming language. He was also the manager of the AT&T ten-year future products development database. His work helped lead to commercial UNIX on computers such as Sun Micros, IBM AIX minis, and Apple computers.

In the 1980's he pioneered the development of PC Desktop Publishing on laser printers and was nominated for three "Awards for Technical Excellence" in 1987 by PC Magazine for PC software products that he designed and developed.

Recently he has developed a theory of Megaverses – actual universes of which our universe is one – with quantum particle-like properties based on the Wheeler-DeWitt equation of Quantum Gravity. He has developed a theory of a baryonic force, which had been conjectured many years ago, and estimated the strength of the force based on discrepancies in measurements of the gravitational constant G. This force, operative in D-dimensional space, can be used to escape from our universe in "uniships" which are the equivalent of the faster-than-light starships proposed in the author's earlier books.

Thus travel to other universes, as well as to other stars is possible.

Blaha also considered the complexified Wheeler-DeWitt equation and showed that its limitation to real-valued coordinates and metrics generated a Cosmological Constant in the Einstein equations.

The author has also recently written a series of books on the serious problems of the United States and their solution as well as a book on the decline of Mankind that will follow from current social and genetic trends in Mankind.

In the past twenty years Dr. Blaha has written over 80 books on a wide range of topics. Some recent major works are: *From Asynchronous Logic to The Standard Model to Superflight to the Stars*, *All the Universe!*, *SuperCivilizations: Civilizations as Superorganisms*, *America's Future: an Islamic Surge, ISIS, al Qaeda, World Epidemics, Ukraine, Russia-China Pact, US Leadership Crisis*, *The Rises and Falls of Man – Destiny – 3000 AD: New Support for a Superorganism MACRO-THEORY of CIVILIZATIONS From CURRENT WORLD TRENDS and NEW Peruvian, Pre-Mayan, Mayan, Anatolian, and Early Egyptian Data, with a Projection to 3000 AD*, and *Mankind in Decline: Genetic Disasters, Human-Animal Hybrids, Overpopulation, Pollution, Global Warming, Food and Water Shortages, Desertification, Poverty, Rising Violence, Genocide, Epidemics, Wars, Leadership Failure.*

He has taught approximately 4,000 students in undergraduate, graduate, and postgraduate corporate education courses primarily in major universities, and large companies and government agencies.

He developed a quantum theory, The Unified SuperStandard Theory (UST), which describes elementary particles in detail without the difficulties of conventional quantum field theory. He found that the internal symmetries of this theory could be exactly derived from an octonion theory called QUeST. He further found that another octonion theory (UTMOST) describes the Megaverse. It can hold QUeST universes such as our own universe. It has an internal symmetry structure which is a superset of the QUeST internal symmetries.

Recently he developed Octonion Cosmology. He replaced it with HyperCosmos theory, which has significantly better features. He developed a fractionalization process for dimensions, particles and symmetry groups. He also described transformation that reduced particles and dimensions to a far more compact form. He also developed a precursor theory ProtoCosmos that leads to the HyperCosmos.

The author showed that space-time and Internal Symmetries can be unified in any of the ten HyperCosmos spaces in their associated HyperUnification spaces. The combined set of HyperUnification spaces enable all HyperCosmos dimensions to be obtained by a General Relativistic transformation from one primordial dimension in the 42 space-time dimension unified HyperUnification space.

The author devel;oped the Cosmos Theory that incorporates ProtoCosmos Theory, HyperCosmos Theory, Limos Theory, Second Kind HyperCosmos Theory and HyperUnification Spaces. He has introduced PseudoFermion wave functions and theory, He has related Cosmos Theory to Regge trajectories of spaces, parton theory, Veneziano amplitudes, Fibonacci numbers and Ramsey numbers. He has calculated an approximation to the difficult R(n,n) Ramsey numbers.

He has developed a Gambol Model that successfully accounts for e-p deep inelastic scattering, fundamental particle resonances, hadron scattering, and the inner structure of particles based on confinement through Casimir forces of ideal gambol gases. The Gambol Planckian Distribution was derived.

He has applied the Gambol Model to particles, universes, and the Cosmos of universes. He showed that the Cosmos may have a distribution of 23 universes corresponding to various Cosmos spaces.

Recently he showed that Cosmos Theory follows from the number of independent asymmetric tensors in a dimension r. He also showed the close parallel between the form of γ-matrices and Cosmos Theory dimension arrays.

Cosmos Theory has now evolved to the point where it may be viewed as deriving all five components based on the Cosmos Fractal Curve. As a result one may view the Cosmos as everlasting – yet "re-created" instantaneously eternally.

www.ingramcontent.com/pod-product-compliance
Lightning Source LLC
LaVergne TN
LVHW071339210926
848486LV00003B/184